The GUN That Made the Twenties ROAR

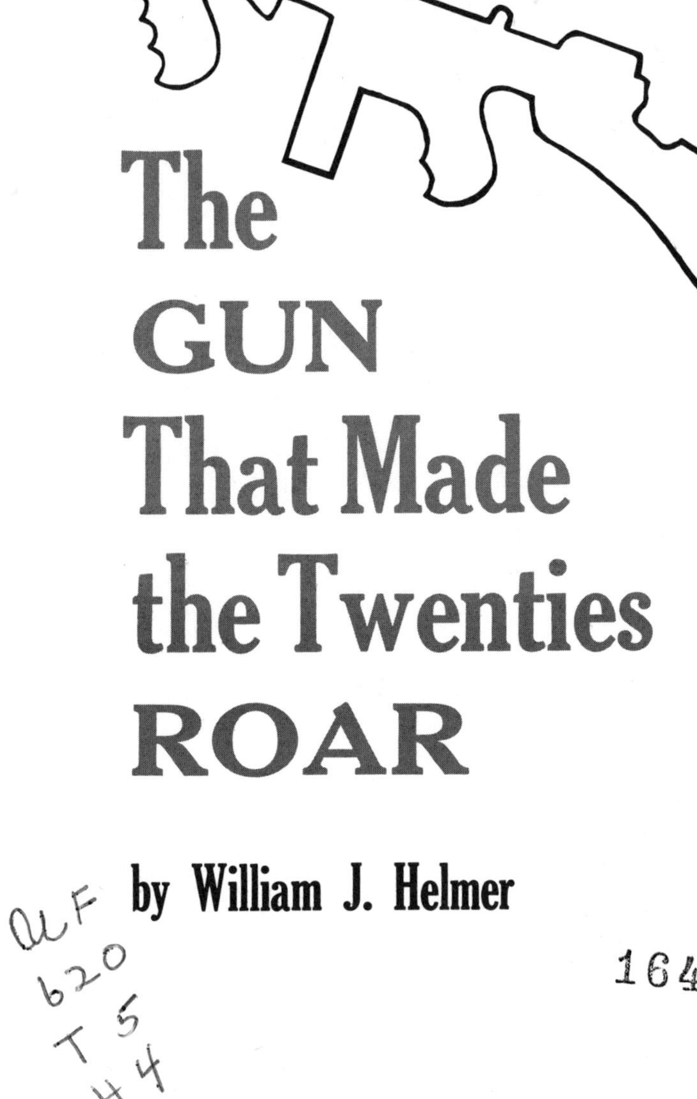

The GUN That Made the Twenties ROAR

by William J. Helmer

THE MACMILLAN COMPANY
COLLIER-MACMILLAN LTD., *LONDON*

FOR
Theodore Eickhoff,
George Goll,
and Oscar Payne

Library of Congress Catalog Card Number: 69-12648

First Printing

The Macmillan Company
Collier-Macmillan Canada Ltd., Toronto, Ontario

Printed in the United States of America

Acknowledgments

This book was conceived in the fall of 1963 as an idea for a master's thesis in history at The University of Texas. The topic seemed original and obscure enough to meet academic requirements, and entertaining enough to hold my attention until I could complete the course work toward an M.A. degree. To dazzle the Graduate School I titled it, on the advice of a professor (a Pulitzer Prize winner who understands such things), *General John T. Thompson and the Gun That Made the Twenties Roar: A Case Study in Culture and Technology.* Five years later the University still had not received *GJTTATGT MTTR:ACSICAT*, nor I an M.A. That my thesis instead grew into a book must be blamed largely on my tolerant supervising professor, Dr. Joe B. Frantz, who helped keep me fed and motivated, and on my three intellectual consultants, Drs. William Goetzmann, David Van Tassel, and Américo Paredes, who contributed helpful suggestions and criticisms. I am grateful to them and to the History Department for their patience (and my recent M.A.).

For the new material in this book I am most deeply indebted to the three men who actually built the first Tommygun, Theodore H. Eickhoff, George E. Goll, and Oscar V. Payne. They supplied many documents and photographs and tirelessly answered letter after letter. Mr. Eickhoff, who retired from civil service work as an Army contracting officer in 1955, still lives in Cleveland. Mr. Goll remained with Thompson's firm, the Auto-Ordnance Corporation, from 1916 through the end of World War II and now lives in retirement in Stroudsburg, Pennsylvania. Mr. Payne joined the Crompton & Knowles Corporation of Worcester, Massachusetts, in 1925, and remained with that company to become director and vice president before his retirement in 1959. I am very sorry to report that Mr.

Payne died May 26, 1967, in New Hampshire, as this book was nearing completion.

I also thank the individuals, companies, and agencies cited in the references, but especially the following: Raymond Koontz, former Auto-Ordnance official and now president of Diebold, Incorporated; George R. Numrich, Jr., of Numrich Arms Corporation; Tom Nelson, author of *The World's Submachine Guns*; Thomas A. Kane, onetime attorney for the Thompson family; Dallas S. Townsend, onetime attorney for Cutts Compensator; General Richard M. Cutts, Jr.; Val Forgett, Jr., of Service Armament Corporation; William B. Edwards, of Gold Rush Gun Shops; Roe S. Clark, Jr., of Savage Arms Corporation; Herbert A. Stewart, formerly with Savage; Don McLean, Normount Armament Company; Stan Friedman, United Press International; Colonel G. B. Jarrett; Colonel George M. Chinn; William B. Ruger; former Auto-Ordnance officials Eugene Daniel Powers and Matthew J. Hall; Tom Pendergast; Ray Bearse; and W. D. Dickinson of H. P. White Laboratory.

Mrs. Evelyn Adams (formerly Evelyn Thompson) of Morristown, New Jersey, was kind enough to loan me General Thompson's personal scrapbook. Dr. J. Bowyer Bell of Harvard suggested the title for Chapter Three and made available his independent research on the submachine gun's role in Irish history. Professor David Hodgson of Ohio State University (who is completing a book on the history of the Lewis gun) looked over the galleys and helped in many other ways.

Others who aided and abetted this project, or suffered through it with me: Pat, Marc, and Jan Helmer; Bob Miller, Joe Steen, and of course Jim Grotte; Mac and Maureen McReynolds; Dwight and Sally Montieth; Glenn and Carmalee Whitehead; Tony Bell and Gilbert Shelton; Leslie Segner, Cornelia Carrier, Doug Allen, Greg Olds, and Bill Day; and Allen, Bill, and Charlie of Dan's No. 2.

Claire Jordan corrected my spelling and everything else, and has earned far more than my gratitude.

W. J. H.

Austin, Texas

Contents

Preface

During the early 1920s Army, Navy, and Marine officials witnessed demonstrations of John Thompson's new hand-held machine gun and marveled that so much fire power could be packed into so small and reliable a weapon. Police officials acclaimed it as a weapon so deadly and versatile that it would either kill or cure the country's gunmen, rioters, and "motorized bandits." In 1926 a *Collier's* writer described the weapon less approvingly: "This Thompson sub-machine gun is nothing less than a diabolical engine of death . . . the paramount example of peace-time barbarism . . . the diabolical acme of human ingenuity in man's effort to devise a mechanical contrivance with which to murder his neighbor." Introduced when it was, the way it was, the notorious "Tommygun" scandalized its inventors, cost a wealthy investor more than a million dollars, and earned itself an enduring place in American folklore as the gangster equivalent of the cowboy's six-shooter.

Unlike the Colt revolver and the Winchester rifle, the Thompson has left its mark on history in bullet holes still too fresh to seem romantic. Most gun collectors scorn it as a modern military man-killer, aesthetic as a hand grenade. Historians have been content with the Thompson's popular image as the gun that made the Twenties roar. Both views are accurate enough. To call a man a Tommygun expert is no great compliment to his sporting skill, and the many sensational machine-gun murders of the Twenties and Thirties have made the "chopper" a symbol of that memorable era.

But the submachine gun's lurid reputation as a gangland murder weapon has largely obscured its significance as a military small arm, its influence on crime and firearms legislation, and its roles in such diverse events as the Irish Rebellion and labor troubles of the Depression. Even less known is the story of its origin and the remarkable man who developed it—Brigadier General John T. Thompson, who lived the last years of his life regretting the gun that bore his name.

1

General John T. Thompson

He was the kind of a man you wanted to work for. He could be so driving and exacting that he'd get on your nerves. Sometimes he expected miracles. But he was always considerate and thoughtful and kind.
—*Theodore H. Eickhoff*

John Taliaferro Thompson devoted much of his life to the United States Army, and most of that Army career to the Ordnance Department, developing or improving the Infantry's small arms, and planning or directing their production. He was an officer and a gentleman in the finest West Point tradition and possessed a blend of personality traits rarely found in combination: strong character and an amiable disposition, imagination and practical ability, a love of weapons and a love of people. His young associates of World War I days remember him as a saint of a soldier whose personal modesty cost him much of the recognition he deserved for his contributions to military ordnance. He seems to have had not an enemy in the world, except the impersonal military kind, and possibly John Browning.

Thompson was born in Newport, Kentucky, on December 31, 1860, the son of Lieutenant Colonel James Thompson and Julia Maria Taliaferro (pronounced "Toliver").[1] On his father's side he was descended from the Thompsons who emigrated from England to Massachusetts in 1630 and later moved to New York. His grandfather was a member of the New York state legislature. His father, sixth in his graduating class at West Point in 1851, distinguished himself as an artillery officer during the Civil War, earning promotions for gallantry in the battles of Glendale and Chickamauga. His mother belonged to the prominent Taliaferro family of Virginia and was related to the Monroes, Madisons, Harrisons, and Taylors of that state.

The war cost James Thompson his health, and in 1869 he retired from active duty to become professor of military science at Indiana University, where he taught off and on until his death in 1880. To his children, John and Frances, an older daughter, he was a strict disciplinarian, a man of strong will and strong character. In manner, however, he was soft-spoken, calm, and philosophical.

John Thompson spent his boyhood on Army posts in Kentucky, Tennessee, Ohio, and California. From his military surroundings and from his father he early acquired a soldierly manner, self-discipline, and a sense of duty; from his mother, a gracious, charitable disposition. Throughout his life he lived by a strict code of personal integrity, forgiving the mistakes of others more easily than his own and concerning himself perhaps too much with living up to his own high ideals. He was an exceptionally happy man nonetheless, and lived life fully and enthusiastically. He was not tall, measuring only five feet seven or eight inches; but he was strongly built, with firm, regular features and intense hazel eyes, and he bore himself with such dignity that he often intimidated strangers. His warm and democratic personality put anyone quickly at ease, however, and those he worked with invariably became his deep and lifelong friends.

By the age of sixteen John Thompson already had decided upon a military career. In 1877 he began preparatory course work at Indiana University, where his father taught, and the following year he entered West Point. He graduated eleventh in the class of 1882, at the age of twenty-one. Before going on active duty he married Juliet Estelle Hagans, the daughter of an Ohio Superior Court judge and a direct descendant of Mayflower Pilgrim Parson John Robinson. A year later their only child was born. He was named Marcellus Hagans Thompson after his maternal grandfather, and like his grandfather James and his own father, he would attend West Point and become an Army officer.

As a second lieutenant, John Thompson's first assignment was with the U.S. Second Artillery at Newport, his place of birth, where he was attached to the same regiment and battery with which his father had served upon graduating from West Point in 1851. From 1882 to 1889 Thompson held routine assignments at Newport and other posts and attended the Army's engineering and artillery schools. In 1890 he was transferred to the Ordnance Department, in which he served through the rank of brigadier general.

The Spanish-American War was a turning point in Thompson's military career. It ended sixteen years of routine and uneventful duty that had poorly suited a man of his energy and imagination and led him into a far more engaging role—that of small-arms specialist. During the war he took charge of supply and logistics; afterwards he devoted himself to modernizing the Army's obsolete rifles, pistols, and service ammunition.

In 1896 Thompson had been detailed to West Point as a senior instructor in ordnance and gunnery. At the outbreak of the war in 1898, the Ordnance Department promoted him to the rank of lieutenant colonel in the Volunteers and appointed him Chief Ordnance Officer under General Shafter in Cuba. During preparations for the Santiago and Puerto Rico campaigns Thompson remained at the Tampa, Florida, ordnance depot trying to untangle the supply problems in which the Army had managed to get itself ensnarled. Supply trains were backed up fifty miles, unable to get in or out; lines of communication had hopelessly broken down. Thompson not only restored order and movement, but, according to one glowing newspaper account, "directed the handling of 18,000 tons of munitions of war without an accident, without the explosion of a single cartridge, and without the discrepancy of one cent in the tremendous item of $4,000,000."[2]

While still at Tampa, Thompson was one day accosted by an excited young second lieutenant, John H. Parker. Parker had learned that fifteen Gatling guns had been consigned to the ordnance depot, but with no orders that would get them to Cuba or into action. Parker wanted the guns; if he could get them to Cuba, he explained to Thompson, he would prevail on General Shafter to organize a special Gatling-gun detachment and prove to the Army the effectiveness of rapid-fire weapons. He also explained to Thompson that he could find no one in authority who would listen to him. Thompson not only listened, but helped Parker assemble an independent command of twelve men with orders to report directly to General Shafter in Cuba. To expedite delivery of the guns, Thompson put Parker in charge of a shipment of reserve ammunition. While the ammunition was being loaded on a transport ship, Parker and his men stealthily lugged the Gatling guns aboard, hid them carefully away, and at the battle of San Juan Hill earned himself a place in Army history as Gatling Gun Parker.[3]

Even before he himself reached Cuba, as Chief Ordnance Officer to General Shafter, Thompson was helplessly aware of the sorry state of American military small arms. The Spanish were equipped with the latest Mauser repeating rifles, Maxim machine guns, and the new smokeless ammunition; American troops were armed mainly with the antique single-shot Springfield rifle designed in the 1860s, plus a few of the newer Krag-Jörgensen bolt actions. Both rifles were supplied only with low-powered black-powder ammunition whose cloud of white smoke gave the enemy a conspicuous target. The Army had no machine guns at all, and only on the personal initiative of the impetuous Lieutenant Parker were Gatling guns brought into action.

The high cost of its victory in the war taught the Army that its small arms were obsolete. In Cuba even the new Krag-Jörgensen rifle could not compete with the Mauser for firepower and accuracy. In the Philippines, the Army's .38-caliber service revolver could not always stop a determined Moro. When Thompson returned to Washington and began arguing for a new line of small arms with more firepower and greater stopping power, the Ordnance Department happily transferred him to the Springfield Armory to see what he could come up with.

After the war Thompson spent seven years at Springfield Armory and the Rock Island Arsenal, studying and developing small arms and serving on various testing boards. At Springfield he supervised the development of a new bolt-action Army rifle, the Model 1903, earning himself the unofficial title, "Father of the Springfield Rifle." The M1903 was based on the Mauser, which Thompson greatly admired, and chambered for a new smokeless, rimless .30-caliber cartridge which soon was refined into the famous .30-'06. The '03 Springfield is still regarded by some as the finest bolt-action military rifle ever built. In slightly modified form and with a telescopic sight, it served as sniper rifle even in World War II, and the .30-'60 cartridge remained Army standard until the nineteen-fifties.

In 1904 Thompson and another officer, Colonel Louis A. La Garde, conducted the Army's extensive tests of handgun ammunition to determine the caliber most suitable for military purposes. The tests consisted of firing carefully aimed shots into live cattle at a Chicago slaughterhouse and into human cadavers obtained from medical schools. Thompson did the shooting and came to the grim conclusion that fat, slow-moving slugs of about .45 caliber were far more destructive than smaller bullets of higher velocity.[4]

In 1907, with the rank of major, Thompson was appointed senior assistant to Chief of Ordnance William Crozier in Washington and put in charge of small-arms design, development, and production. He organized the Ordnance Department's Small Arms Division to handle this job, and also worked out a plan for the prompt equipping of a 500,000-man emergency force, with supply arrangements carefully worked out to avoid the chaos that had marked the Spanish-American War. In Crozier's absence from Washington, Thompson served as Acting Chief of Ordnance.

At Springfield, Rock Island, and Washington, Thompson worked closely with the Infantry and Cavalry equipment boards in the development of small arms and accessories. Articles in Army newspapers credit him as the first Ordnance Department officer to perceive the merits of automatic pistols and rifles and to urge their adoption. According to one published account, Thompson's "was the mind which first grasped the possibilities of an automatic pistol, and his efforts in this direction are probably more responsible than anything else for the working out of the present service Colt's automatic pistol [the Model of 1911], a weapon believed to be superior to any other of its class."[5] Thompson also was instrumental in the development and adoption of the .45-caliber rimless cartridge. When he was promoted to colonel in 1913 a service publication paid him the following tribute.

> No man of his grade in the Army of the United States has performed more distinguished and valuable service than Colonel Thompson. This is natural, because he is an officer of exceptional ability. He has the type of mind in which the constructive faculty is highly developed; that is, he has "that power of intellect by which the soul groups knowledge into systems, scientific, artistic, and practical." Moreover he has the practical imaginative quality by which he is able to project his thoughts into the future and resolve an occasion yet to arrive. . . .
>
> The friends of Colonel Thompson are very numerous. He has that admirable quality in an administrative officer of putting at ease every one who comes in contact with him. At the same time he always knows his own mind, is quick to make decisions, expeditious in his actions, and of unswerving truthfulness to himself and his obligations.
>
> Such a man is a very great credit to the soil which produced him and to the Army in which he serves.[6]

John Thompson probably contributed as much or more than any other individual to modernizing infantry ordnance during the period from 1900 through World War I. Even so, his achievements fell far short of his ambitions. Thompson's experiences in Cuba made him a strong advocate of machine guns, automatic rifles, and "firepower," but the Ordnance Department, content with its new Springfield rifle and Colt pistol, showed little more than dutiful interest in the automatic principle, and a good deal of indifference. Much of this opposition was simple inertia on the part of a few musket-era generals who held tenaciously to the old "one shot, one man" idea. An automatic rifle, they contended, would only tempt a soldier to spray his bullets wildly, which would waste ammunition, create supply problems, and ruin military marksmanship. Opposition to the machine gun was even less enlightened: refusal to concede any tactical value to such a weapon.[7]

Although the inventors who developed the first successful rapid-fire guns and pioneered the automatic principle were mostly Americans—Gatling, Maxim, Browning, Lewis, Borchardt, Benét, Hotchkiss—the United States lagged far behind other major countries in adopting such weapons or even recognizing their potential. Richard Gatling developed the first practical hand-operated "machine" gun during the Civil War and offered it to the Union. But because Gatling was himself a Southerner and was believed to be sympathetic to the South, the Army's ordnance officials rejected the gun, so the story goes, out of suspicion that it might be some kind of Rebel trick. Toward the end of the war General Benjamin Butler personally ordered the purchase of twelve Gatling guns and demonstrated their effectiveness during the siege of Petersburg, Virginia. After the war the Army adopted the gun and acquired a small number for service on the frontier. But on one historic day in 1876, General George Custer, because of rough terrain and a certain amount of indifference, left his Gatling guns at home.

Because the Gatling gun resembled a field piece, Army tacticians tended to regard it as some new kind of inferior cannon which fired its grapeshot one piece at a time.[8] Even after Hiram Maxim invented the fully-automatic water-cooled machine gun, and its deadliness had been proved in the Boer and Russo-Japanese wars, the Army persisted in writing it off as little more than an ordnance novelty that burned ammunition and boiled like a steam engine. As late as 1911, Army drill regulations treated the machine gun under the heading "Miscellaneous" and called it a "weapon of emergency," valuable only at "infrequent" periods. "Firepower alone," the Army contended, "cannot be depended upon to stop an attack."[9]

As an advocate of firepower and machine weapons, Thompson was something of an Ordnance Department radical. In papers and in talks he predicted the eventual adoption of the automatic principle in all military ordnance. He argued that mechanization was revolutionizing warfare and that land battles of the future would be won or lost on the basis of mobility and close-range offensive firepower. Traditional infantry tactics placed importance on numerical superiority, which of course implied superior firepower. But Thompson believed that firepower could be either a substitute for or a complement to manpower, that one infantryman with an automatic weapon was equal to several with slower-firing rifles, and that one well-placed machine gun could hold back a company.[10]

Lacking the temperament of a Billy Mitchell or a Gatling Gun Parker, Thompson did not crusade on behalf of automatic weapons. His personal interests centered on rifles, not machine guns, and it was primarily with a view to increasing the firepower of the service rifle that he concerned himself with automatic weapons at all. About 1909 he induced the Ordnance Department to conduct an extensive study of both American and foreign automatic rifles, but nothing came of it. The department found them generally too heavy for infantry purposes, and too complex to function reliably under combat conditions.

That the Ordnance Department made little effort to develop a suitable automatic rifle could be attributed partly to the same conservatism that caused the Army to scorn the machine gun. Neither weapon fit neatly into existing tactical theory. But other factors also delayed acceptance of the automatic gun, especially in the United States. After the Spanish-American War, Congress cut military budgets to the bone, so the Ordnance Department complained, leaving no funds for research and development. Nor, in peacetime, did the Army feel any urgent need to develop new weapons or spend money improving them. Military policy was, in effect, to let the private munitions industry and foreign countries take care of inventing and refining ordnance, and then harvest the fruits. In practice this policy had serious drawbacks. It drove this country's most progressive arms inventors to Europe, where they tailored their work to the specifications of foreign firms and foreign governments—Hiram Maxim in England, Hugo Borchardt in Germany, John Browning and I. N. Lewis in Belgium, Benjamin Hotchkiss and Laurence Benét in France. It also meant that the United States Army could postpone indefinitely the adoption of existing automatic guns, because none met domestic military specifications. When the United States went to war in 1917 it had on hand some 440 Model 1895 Browning and Model 1904 Maxim machine guns, and none on order. Germany had gone to war three years earlier with 12,500 late-model Maxims and some 50,000 on order.[11]

Even after the outbreak of war in Europe the Army continued to resist automatic weapons on principle, reject them on technical grounds, and ignore them for lack of any imminent military need. When the German victories in Belgium and France failed to arouse the Ordnance Department or to loosen Congressional purse strings, John Thompson abandoned his efforts to interest the Army in automatic rifles and began making plans to develop one privately. He believed he could design a light, dependable semiautomatic shoulder rifle whose performance would win the heart of even the most hidebound general in the Army.[12]

In November 1914, Thompson surprised his Ordnance Department colleagues by retiring from the Army to become chief consulting engineer for the Remington Arms Corporation. His Ordnance Department work had established him as a small-arms expert, and Remington offered him a challenging job: designing the world's largest rifle factory at Eddystone, Pennsylvania, supervising its construction, and directing the manufacture of rifles for the British army. Thompson got the British .303 Enfield into production in October 1915, two months ahead of schedule, and by the middle of 1916 Remington was mass producing the rifles at the rate of two thousand a day. By this time Remington also had received large orders from Russia to manufacture the 7.62mm Mosin-Nagant. To handle the Russian contract, Thompson supervised the building of another large rifle factory, at Bridgeport, Connecticut.

Manufacture of the British rifles had gone smoothly from the start, but Remington encountered one problem after another with the Russian rifles, and with the Russians. The blueprints, gauges, and master models supplied by the Russian government all differed from one another, and it was months before the Russians would approve certain Remington modifications of the rifle that would permit the mass production of interchangeable parts. To complicate matters further, the Czar supplied Remington with fifteen hundred official government rifle inspectors, who did little to increase plant efficiency.[13] Getting the British and Russian rifles into production left Thompson little time to work on his own projects, but it gave him the manufacturing experience that would make him one of the country's most valuable ordnance experts when the United States entered the war in 1917.

Thompson joined Remington ostensibly for the same reason many other professional soldiers joined private companies in 1914 and 1915—to participate in the war effort. The war in Europe affected the country's munitions and other industries long before it affected the Army, whose continuing inactivity created a pool of restless officers. The United States was officially and militarily neutral, but in the burgeoning war industries one still could do his part for the Allied cause. Also, one could make a good deal more money.

John Thompson, however, left the Army primarily because his Ordnance Department work did not afford him the opportunity to develop an automatic rifle. To a subordinate, later his employee, he confided that his interest in Remington was secondary to the rifle project that he was working on late at night at his country home near Chester, Pennsylvania. Once he had a promising design, he would develop it with private capital and submit it to the Ordnance Department as a civilian inventor.[14]

The chief technical problem confronting Thompson was working out a simple, dependable, self-operating breech mechanism—the heart of any automatic weapon. The Gatling and most other early rapid-fire guns were not, strictly speaking, self-operating. The gunner turned a crank and a geared mechanism loaded, fired, and reloaded a cluster of barrels as they rotated about a central axis. In 1884 Hiram Maxim developed a truly automatic machine gun that operated on the force of its own recoil. With each shot the barrel moved back and forth in its housing, setting in motion the machinery that unlocked the bolt, opened it to eject the spent cartridge, closed and locked it on a fresh cartridge fed from a belt, and cocked and released the hammer. In 1890 John Browning invented the first practical machine gun with a breech mechanism actuated by gas pressure; a tiny hole drilled in the barrel diverted a portion of the expanding gases into a chamber to operate the gun's action by means of a hinged flap or a piston.

Recoil actuation became the most popular system for heavy and medium machine guns, as did gas actuation for light machine guns and automatic rifles. But both systems, especially in their early forms, used numerous moving parts, which made machine weapons heavy and relatively unreliable.

In addition to recoil and gas actuation, a third system, incorporating no breech lock whatever, found wide application in automatic arms. Called "blowback," it utilized rearward gas pressure to literally blow the cartridge case back against the bolt—with enough force to drive it open, compressing a coil spring, ejecting the case, and recocking the hammer. As the bolt reclosed it picked up and chambered a fresh cartridge. The blowback system was simple and therefore reliable. But because it depended mainly on the inertia of the bolt to keep the breech closed during the moment of highest chamber pressure, it could be used only with low-powered pistol ammunition. With rifle ammunition an unlocked bolt would blow open prematurely, resulting in ruptured cartridge cases and the explosive escape of gases.

During 1914 and 1915 Thompson, in what little spare time he had, studied the existing types of automatic breech mechanisms. He also searched the Patent Office files and uncovered a pending application that seemed ideally suited to his needs. It was a modified blowback system with a self-actuating lock. Instead of a train of levers, gears, and rods linked to a piston or to a floating barrel, it employed a simple metal wedge that delayed the opening of the bolt until the chamber pressure had fallen to a safe level, then allowed the residual pressure to blow back the bolt as in an ordinary .22-caliber self-loading rifle. The patent had been applied for in 1913 by John Blish of Brookline, Massachusetts, and was granted in March 1915, as a "Breech Closure for Firearms," No. 1,131,319.

John Bell Blish[15] was born in Indiana in 1860, the same year as John Thompson. He entered the Naval Academy at Annapolis at the age of fifteen and later served as a Navy mathematician, a coastal surveyor, and executive officer on the cruisers *Niagara* and *Vicksburg* during the Spanish-American War. He retired in 1905 with the rank of commander, returned briefly to active duty during World War I, and died in 1921.[16]

Both as a Navy officer and as a mathematician Blish long had been curious about a phenomenon he once had observed during an inspection of a naval gun installation: when firing light charges the guns had a tendency to unscrew their breech blocks and fly open, endangering their crews; yet the same guns, when loaded with heavy charges, remained closed and functioned perfectly. After pondering this and conducting some experiments, Blish came to the conclusion that metals had a tendency to adhere to each other *under pressure* with a force much greater than simple friction. He called this "the principle of metallic adhesion"; later it became known as the Blish Principle.[17]

It soon occurred to Blish that this principle might lend itself to a self-locking, self-opening breech mechanism for small arms. The purpose of a lock was to keep the breech sealed tightly as long as the chamber pressure was dangerously high. If a lock could be designed to do this, but release itself automatically under reduced pressure, then the breech mechanism could operate by simple blowback regardless of the power of the charge.

After retiring from the Navy Blish began work on both a rifle and a pistol mechanism that would demonstrate his proposed locking system. He applied for a patent in 1913 and in 1915 submitted his invention to the Special Capehart Board of Naval Ordnance. The Board made a valorous effort to fathom Blish's mysterious "adhesion" principle, without great success.

> On the subject of adhesion, the Board finds the literature scant. In every day life this force manifests itself chiefly in the adherence of liquids to solids, especially when such liquids "wet" the solids, for example, water and glass. It may be noted, however, that the efficiency of all glues and cements depends largely upon the force of adhesion.

> Occasionally it manifests itself in two solids such as when two plates of polished glass adhere with sufficient force to render it difficult to separate them without breaking them.[18]

But as the experimental mechanisms seemed to work quite well in tests, the Board decided to accept Blish's explanation, reporting:

> Adhesion is a molecular force which binds together the surface of molecules of two bodies at the common surface of contact. . . . If two bodies are pressed together, the pressure increases the intimacy of contact, especially if the pressure be great, and thus indirectly causes the surfaces to adhere. If the pressure be gradually reduced the contact becomes less perfect and a point is reached where the contact is not sufficiently perfect for the surface to adhere—adhesion ceases and the surfaces are free to move. . . .
>
> The rifle mechanism was especially interesting for several reasons, *viz*: (1) The wedge angle was considerably less than the angle of friction. (2) Sufficient pressure remained in the bore when the wedge opened, for the empty cartridge case to be blown to the rear, carrying the bolt . . . before it, with sufficient velocity to cock the firing pin when the bolt brought up against the rear buffer, and causing the empty cartridge case to rebound and to be ejected to the front.[19]

Blish's experimental guns operated by "delayed blowback," and the delay, or momentary locking effect, was achieved by requiring the bolt to work against an obstructing metal wedge riding in steeply inclined slots in the receiver. Under heavy pressure the wedge would adhere tightly in its slots and prevent the bolt from moving backward, but under low pressure it would slip down and out of the way, permitting the bolt to move freely. Blish's models were not automatic insofar as they had no magazines and each shot had to be chambered by hand. But they unlocked, opened, ejected, closed, and locked again by means of only two moving parts—the bolt and the wedge—and it was in this feature that John Thompson saw the makings of a simple, dependable, lightweight automatic rifle.

Some time in 1915 Thompson contacted Blish and outlined his plans for developing a rifle. Blish was enthusiastic and agreed to permit the use of his patent in return for a block of stock in the new arms company that Thompson intended to organize as soon as the financing could be arranged. Both men were confident that a practical high-powered military rifle could be designed around the Blish system and that the Ordnance Department, despite its prejudices, would not be able to turn down such a weapon.

Thompson's decision to design a new gun independently and develop it privately was a high-stakes gamble, as he doubtless knew. The United States, unlike England, France, Germany, and other countries, has never had a munitions firm (like Vickers-Armstrong, Schneider-Creusot, or Krupp) working hand in hand with the government to supply the country's armament needs. Instead the United States has operated its own military arsenals—the design, development, and manufacture of military armaments being the responsibility of the Ordnance Department and the Navy's Bureau of Ordnance. In obsolete theory, these arsenals are adequate to supply the military's peacetime armament needs, and operate independently of the country's privately owned munitions industry. However, private firms and individuals are invited to submit their inventions for government consideration, and in time of war the government routinely contracts with commercial manufacturers to supplement its own arsenal production.

Under this system, a private arms firm could expect no government subsidy of its research and development. Profits on an item of military ordnance depended entirely on the good fortune of having it officially adopted by the United States, as in the case of the Colt-Browning pistol, or marketing it to a foreign government, as in the case of the Lewis light machine gun. If a company could find no market here or abroad, it had to absorb the high cost of developing a new weapon and try another one. Colt's company could stand to lose such gambles now and then—but not a new, highly specialized company such as that proposed by Thompson. Moreover, Thompson would have to compete with old, established firms, like Colt's, which over the years had developed a close if unofficial relationship with the military. Colt's produced guns, DuPont powder, and Bethlehem armorplate; and together these three companies constituted a kind of American Krupp works.

On the other hand, military adoption of a privately developed weapon either here or abroad, especially in wartime, meant enormous profits, and the United States government had always preserved this incentive to invent. Instead of buying patent rights, it paid royalties; and it jealously guarded the right of Americans to export arms. In 1793 Thomas Jefferson answered a British protest against arms shipments to the French with a statement that became an item of American foreign policy:

> Our citizens have always been free to make, vend, and export arms. It is the constant occupations and livelihood of some of them. To suppress their callings . . . because a war exists in foreign and distant countries, in which we have no concern, would scarcely be expected. It would be hard in principle and impossible in practice. The law of nations, therefore, respecting the rights of those at peace, does not require from them such an internal disarrangement in their occupations. It is satisfied with the external penalty . . . of confiscation of such portions of these arms as shall fall into the hands of the belligerent powers on their way to the ports of their enemies. To this penalty our citizens are warned that they will be abandoned, and, that even private contraventions may work no inequality between the parties at war, the benefit of them will be left equally free and open to all.[20]

With only rare exceptions this policy was followed until 1917. It guaranteed a healthy domestic arms industry which could be mobilized quickly in time of national emergency, thus sparing the government the cost either of subsidizing the industry during peacetime or of building huge arsenals that most of the time would lie idle.

This convenient arrangement ended with America's entry into the World War and world politics. After the war the country's munitions makers found themselves with a domestic arms market that for lack of military spending was in a state of virtual collapse, and also with the only promising foreign markets ruled out by an increasing number of politically necessary export restrictions. But in 1914 the war in Europe was expanding and intensifying, and the time seemed more than ripe for introducing a revolutionary rifle that would increase tenfold or more the firepower of an infantry soldier. Thompson, who had many friends in high military, political, and financial circles, had little trouble raising the necessary funds.

A business associate once described Thomas Fortune Ryan as "the most adroit, suave, and noiseless man American finance has ever known."[21] Ryan's business rivals described him less appreciatively as the last of the robber barons, a ruthless manipulator who stripped and looted corporations through the device of the holding company. Less colorful than many of his Gilded Age contemporaries, but no less shrewd, he became known as "the gray ghost of Wall Street."[22]

Born on a Virginia farm in 1851 to Irish immigrant parents, orphaned and penniless at fourteen, Ryan started as a Wall Street messenger boy and worked his way to the top in the Horatio Alger tradition, if not always by sterling Alger methods. He carved his first wealth out of public transportation in New York City in the 1880s, and left his overcapitalized corporations in ruins when he pulled out suddenly in 1906. With William Whitney he organized the American Tobacco Company, which he turned into a $250,000,000 monopoly through a series of forced mergers. At the invitation of King Leopold, Ryan went to the Belgian Congo, developed Belgian properties, and received in return a quarter interest in the Congo's copper, gold, and diamond mines. The so-called holding company was an idea that Ryan pioneered; another of his cunning financial maneuvers involved gaining control of his sources of credit. Because reputable Wall Street banking firms did not trust the Ryan-Whitney syndicate and sometimes balked at financing its schemes, Ryan quietly acquired a controlling interest in several banks and installed boards of directors who did as they were told, banking laws notwithstanding. After 1910 Ryan retired from the active management of his far-flung enterprises—surrounded by investigations, banking scandals, charges of political bribery, and wealth approaching $200,000,000.[23]

From Ryan, Thompson obtained the capital to develop his automatic rifle. The two men had met each other through Thompson's son, Marcellus, who married Dorothy Harvey, the daughter of Colonel George Brinton McClellan Harvey, Ryan's personal friend and business associate. Harvey was himself quite a colorful character—either a great journalist, statesman, and patriot or a thoroughgoing scoundrel, depending on who was describing him, and when. A lanky Vermonter, born in 1864, Harvey worked as a reporter for newspapers in Vermont, Chicago, and New York until 1891, when Joseph Pulitzer appointed him managing editor of the New York *World.* Over the next thirty-five years he edited the *North American Review, Harper's Weekly, Harvey's Weekly*, and the *Washington Post.* Notorious for his skill at satirizing public figures and policies, Harvey used his journalistic influence to work his way into the highest levels of politics and finance, where he became friendly with Thomas Fortune Ryan. According to his detractors, Harvey devoted his publications to whitewashing Ryan's and Whitney's Wall Street villainy in return for a cut of the take, and toiled in smoke-filled rooms to make and break presidential candidates.[24]

Harvey did engineer the nomination of Warren G. Harding in the original smoke-filled room—his own—at the Hotel Blackstone in Chicago in 1920, and for his trouble received the post of ambassador to Great Britain. His appointment at first dismayed the British,[25] and Harvey did little to improve his image by arriving at the Court of St. James for his formal presentation foppishly attired in satin knee breeches, silk stockings, and silver-buckled shoes. He also made a spectacle of himself by racing about London at the wheel of a flivver instead of using the customary chauffeured limousine. Harvey's motoring habits inspired one journalist to poetry.

HARVEY AND LIZZIE AT THE COURT OF ST. JYMES[26]

I'm ambassador, sir, to the Court of St. Jymes—
To the seats of the mighty I've soared;
To show 'em a diplomat up with the times
I skip around town in a Ford!

Down gay Piccadilly the bally old boys
Admit that I've certainly scored:
They say when their ears catch anelluva noise
" 'Tis Ambassador George in his Ford!"

All precedent into the discard I've hurled;
By conventions I'm terribly bored;

I find that an envoy who'd rattle the world
Can best do the thing with a Ford.

Honk! Honk! Clear the way, there, I'm putting on speed!
I'll soon have the court very dizzy.
Old England must reckon from now on, indeed,
Not only with George, but with Lizzie.

Despite his controversial background and playful habits, Harvey quickly won the hearts of the British (and made himself even more enemies at home) by frankly minimizing the United States' role in winning the Great War. In his first formal speech he told a London audience (referring to American troops):

> We sent them solely to save the United States of America, and most reluctantly and laggardly at that. We were not too proud to fight, whatever that may mean. We were afraid not to fight. . . . And so we came along toward the end and helped you and your Allies to shorten the war. That is all we did, and all we claim to have done.[27]

After a year in London, Harvey was applauded by *The Spectator* as a man of candor and good will whose skillful diplomacy had done much to resolve the postwar differences between the United States and Great Britain.[28] Shortly after the death of President Harding in 1923 Harvey resigned his post as ambassador and returned to the United States. He returned to editorial work for a time, but retired in 1926 to look after his health and to write the biography of his long-time friend Henry Clay Frick, the steel magnate. He died of a heart attack at his summer home in New Hampshire on August 20, 1928.

Despite their radically different interests and personalities, John Thompson and George Harvey became good personal friends. When Harvey heard Thompson's plan to develop an automatic rifle, he not only offered to invest in the business but helped Thompson sell the idea to Thomas Fortune Ryan. Partly as a favor to Harvey, Ryan agreed to bankroll the venture. In August 1916 the Auto-Ordnance Corporation of New York was formed, Ryan supplying all development and operating funds in return for a controlling interest in the company. Of the 40,000 shares of stock authorized, about 18,000 were issued to the Ryan interests, 1,500 went to John Blish for the use of his patent, and about 10,000 were divided among the Thompsons and Harveys.[29]

Ryan's interest in Thompson's undertaking was largely altruistic. He was extremely wealthy, retired; he had never concerned himself with munitions in any way; and he never permitted his name to be used publicly in connection either with the Auto-Ordnance Corporation or with Thompson guns. Considering the magnitude of Ryan's other holdings, the profit motive could not have been overwhelming. Some twenty years later a magazine article attributed Ryan's interest in Auto-Ordnance to wartime patriotism. Ryan's patriotism, however, was not ordinarily of the philanthropic sort—except possibly toward Ireland. Partly credible rumors had it that Ryan money helped finance the Irish Rebellion, and a gunrunning incident in 1921 made it appear that "Ryan men" had an interest in Thompson guns that was not purely commercial.

The Auto-Ordnance Corporation began operations in the summer of 1916 as virtually a secret organization. It had no offices, no property, no assets, no bank account, and no money. Thomas Fortune Ryan supplied all operating funds in the form of personal checks made out to Thompson's chief engineer, who conducted all Auto-Ordnance business in his own name. The purpose of this arrangement was to conceal the nature of Thompson's project and even the name of the company. For reasons of both commercial and military security, Thompson wanted his new rifle to come as a surprise.[30]

The chief engineer was Theodore H. Eickhoff of Indianapolis, thirty years old.[31] Eickhoff had graduated from Purdue University in 1908 and had taken a civil service job in Washington as an electrical and mechanical draftsman in the Ordnance Department. A year later he was transferred to the Small Arms Division as an assistant to John Thompson, then a major. From 1909 to 1914 Eickhoff worked closely with Thompson, studying commercial and military rifles and testing the Colt and Savage automatic pistols, which were then competing for adoption. It was Eickhoff's duty to witness the tests, record the results, and make reports. In 1911 Eickhoff delivered a refined model of each pistol to the commanding officer of the Springfield Armory for the final tests. The results favored the Browning-designed Colt, and its adoption was approved that year by the Secretary of War.

In the fall of 1914 Thompson retired from the Army with the rank of colonel and joined Remington. The following spring Eickhoff, who had also grown restless in the Ordnance Department, quit his civil service job to look for work in private industry. He found nothing that suited him, however. After several months of job hunting, he returned to Indiana to help run the family cider mill.

After more than a year of uninteresting temporary jobs, including cider making, Eickhoff was beginning to regret that he had left the Ordnance Department. Then, in July 1916, he received a telegram from Colonel Thompson. To his great excitement, it invited him to Thompson's home in Pennsylvania to discuss his possible employment on a special ordnance project. Eickhoff, dressed in his best suit and a new straw boater, took the next train to Chester and was met at the railroad station by Colonel Thompson and his chauffeur.

During the drive to his home, Thompson talked mostly about the Ordnance Department—its skeptical attitude toward automatic rifles and its paucity of funds for research. He continued the discussion at home, outlining his plan to design and build a military weapon around a newly patented breech-locking mechanism. He explained the theory of the Blish system in some detail and told of his arrangements with Thomas Fortune Ryan to form the Auto-Ordnance Corporation. Finally, he asked Eickhoff if he would like to be the company's chief engineer. The offer flattered Eickhoff, and also amazed him, for he had never regarded himself as a small-arms expert. But the opportunity was not one that he could pass up. He accepted eagerly, hopeful that even if he was no expert at that moment, he soon could become one.

At this time the only other employee of Auto-Ordnance was the young man who drove Thompson's car. George E. Goll, twenty-seven years old, from Paradise Valley in eastern Pennsylvania, had been working as a fireman on the Philadelphia & Reading Railroad.[32] During a particularly slack period Goll found himself running short of money and started looking for a part-time job as an automobile mechanic. He found no mechanic's job, but one car dealer he applied to suggested he talk to Colonel Thompson. Thompson had just purchased a new Winton Six and was looking for a driver. Goll not only got the job, but impressed Thompson as being a bright young man with a good deal of mechanical aptitude. Thompson asked him if he would like to learn small-arms designing as an assistant on the rifle project. Like Eickhoff, Goll accepted enthusiastically.

It was Thompson's policy to judge men primarily on their character and to place more importance on their talent and ambition than on their experience. He especially liked young men who were not yet set in their ways of doing things and who were motivated by a desire to advance themselves professionally. Neither Eickhoff nor Goll had the formal qualifications to design a completely new firearm. But Thompson reckoned, and reckoned correctly, that they had enough youthful enthusiasm to do it anyway. Thompson once summed up his executive philosophy to a Philadelphia reporter interviewing him on his Ordnance Department work: "My principle has been to select big, broad, efficient men, and to put them to work at something they know how to do, or at some similar line of work. Say to them, 'There's your job. Go to it in your own way. Stand on your head if you want to in getting it done, but get it done.' "[33] He emphasized the importance of the "human factor" in engineering, of treating subordinates courteously and fairly, and placing confidence in them.

"He was the kind of a man you wanted to work for," Eickhoff recalls. "He could be so driving and exacting that he'd get on your nerves. Sometimes he expected miracles. But he was always considerate and thoughtful and kind. At the Ordnance Department I was a young man, just out of college, and one day when my mother happened to be visiting me at work he met her in the hall. I was nothing to him but an employee, but he stopped and chatted with her for a long time. She was so impressed, and so was I. But if you did something he didn't like, he could surely take you to task for it. He could give you a mental pinch."[34]

One such "pinch" that Eickhoff remembers vividly occurred when he walked into Thompson's office carrying a rifle barrel. It was merely an empty, unattached barrel, and Eickhoff allowed it to point around the room as he carried it over to a table. "Don't ever do that," Thompson said sternly. "Don't *ever* allow the barrel of a firearm to cross a man's body." He proceeded to lecture Eickhoff at some length on the importance of developing good habits with regard to handling firearms, and the disrespect for human life implied in allowing even a part of a deadly weapon to point at another person.[35]

In August 1916 Theodore Eickhoff moved into a spare room in the Thompson home and became "a member of the family." The "family" at this time included the Colonel and Mrs. Thompson, a Negro servant named Willy Smith, who had been Thompson's orderly in the service, and George Goll. In another room, equipped with a drafting board and a few tools, work began on the Thompson "Autorifle."[36]

To demonstrate his locking system, John Blish had constructed a small single-shot handgun whose shape resembled that of a .38-caliber revolver. Between the hammer and the cartridge chamber it had a small metal wedge riding in two slots inclined downward and slightly toward the rear. The wedge, held at the top of its track by spring pressure, acted as a breech block. Owing to Blish's principle of metallic adhesion, it adhered in place at the moment of firing, sealing the cartridge in the chamber. Then, as the chamber pressure receded, it became unstuck and slipped down, exposing the chamber and permitting the residual chamber pressure to blow the empty case straight back. A curved plate behind the chamber deflected it away from the face of the firer.

Thompson, Eickhoff, and Goll took this experimental pistol into the woods behind the house and fired it repeatedly to observe the action. Eickhoff then drew up plans for a test mechanism, using a Springfield .30-'06 rifle barrel, a receiver, and a Blish lock whose sliding wedge could be adjusted to work at different angles.

To build the first test mechanism, Thompson engaged the Warner & Swasey Company, a large machine-tool firm in Cleveland. W. R. Warner and Ambrose Swasey both had become personal friends of Thompson as a result of their contract work for the Ordnance Department. Thompson had a high regard for the company and also for its manager, Frank Scott, who later became chairman of the Army's General Munitions Board and president of the War Industries Board. Thompson discussed his project with the Warner & Swasey officials and quickly aroused their enthusiasm. They assigned several of their best machinists and engineers to work closely with Eickhoff and provided a testing room in the basement of their plant at 5809 Carnegie Avenue.

Around the end of November 1916 Warner & Swasey completed the testing mechanism and Eickhoff began making trips to Cleveland to conduct experiments with it. As the weeks went by, Eickhoff's visits to Cleveland became more and more extended, and in February 1917 he moved there to open a small office, hire an independent staff, and carry on Auto-Ordnance work full time. Two months later, in April, the United States entered the war. Government work orders began pouring into Warner & Swasey, tying up the machine shop and delaying the construction of new test mechanisms for Auto-Ordnance. This presented Eickhoff with a dilemma. He was under direct orders from Thompson to farm out all machine-shop work; he was also receiving almost daily reminders from Thompson that with the United States in the war, time was "of the essence." When the war resulted in Thompson's return to active duty in Washington, he could no longer keep close tabs on the Cleveland office. Eickhoff took advantage of this to set up a shop "on the sly," installing machine tools that were begged, borrowed, or bought with the leftovers from a few padded expense vouchers.[37] When told later what had taken place, Thompson allowed that the idea was probably a good one, under the circumstances.

With a shop and several full-time machinists, work on the Autorifle resumed at an intensive pace in the late spring of 1917. One problem after another kept cropping up, however, and progress was worse than slow. The Blish lock would work properly a few times and then jam, and it developed that even the hardened nickel-steel bearing surfaces of the sliding wedge were abrading and then freezing under the back pressure of high-powered rifle cartridges. Eickhoff began looking for a better metal for the wedge and tried several kinds without success. Then a metallurgical engineer for Warner & Swasey, William Burger, suggested making the wedge out of titanium-aluminum bronze, a special alloy developed by a metallurgical firm in Niagara Falls, New York.

The bronze alloy lock solved the problem of surface abrasion, but the system still refused to function dependably, owing to a tendency of the fired cartridge cases to expand slightly and stick in the chamber. Eventually it was observed that the only cartridges ejecting properly were those that had been handled enough to pick up a slight amount of oil or grease. Further experiments proved that to function consistently in the Blish breech mechanism, cartridges required lubrication.

Eickhoff came to this conclusion in May 1917, and with considerable disappointment. For military purposes, lubricated cartridges were not desirable, as they would have a tendency to pick up dirt, lint, and other foreign matter that might cause jamming. Nor was it desirable to build a lubricating system into the rifle itself, as this would further complicate the mechanism. And in either case there would be the problem of the lubricant, especially if it was wax or paraffin, accumulating in the chamber.

To make matters even worse, the Blish lock's unforeseen problems had caused an important gamble not to pay off. Because time was "of the essence," Eickhoff had designed a complete prototype Blish-system rifle back in the fall of 1916, taking the chance that it would require no more than minor modifications. By January 1917 Warner & Swasey had built a brass model of the rifle. By summer the finishing touches were being added to the fireable steel version, which was designated the Thompson Autorifle, Model 1. By the time the rifle was finished, however, it was clear the effort was largely wasted, and during its first trial, in August, it blew up.

Confronted by one setback after another, Eickhoff began losing hope that the Blish system could be developed into a practical rifle or anything else. The problem with the Model 1 Autorifle could be corrected; it had burst because the sliding wedge was independent of the bolt and had not yet seated itself and locked the bolt at the moment the cartridge fired. But there remained the problem of cartridge lubrication, which amounted to an absolutely undesirable necessity.

Eickhoff tackled both problems at once, despite his growing suspicions of the entire project. He authorized the construction of a second Autorifle, the Model 2, with a modified lock that would not permit the bolt to close and fire the cartridge until the lock was completely seated. He also put his assistants to work on internal lubrication systems. Meanwhile, he began a series of experiments with a new testing mechanism to determine the "coefficient of extraction" for different types of cartridges, hoping to find a cartridge that would function "dry" with the Blish lock.

The coefficient of extraction was merely the ratio of the bore area to the surface area of the cartridge, and the tests showed the cartridge with the lowest coefficient was the .30-'06. Its slender shape provided a large surface area in contact with the chamber walls, and it had a small bore, which represented the effective thrust area on the base of the cartridge. The greater the surface area compared to the thrust area, the greater the tendency to stick. The cartridge with the highest coefficient of extraction was the short, fat .45-caliber Colt automatic pistol round. Tests showed it to work perfectly in the Blish system without any kind of lubrication.

In his earlier reports to Thompson, Eickhoff had always managed to sound optimistic. He could even call the Model 1 Autorifle "a success in a way" because it illustrated beautifully what not to do next time. But now the results of the extraction study seemed to doom the entire project, which required an unlubricated *rifle* cartridge. Discouraged and dreading what he had to do next, Eickhoff traveled to Washington in September 1917 to inform Thompson that the only service cartridge apparently suited to the Blish-lock rifle was the one least desirable—the Colt pistol cartridge.

Eickhoff cannot recall whether he broke the news to Thompson in his study at home or in his office at the Ordnance Department. But he remembers that Thompson was sitting behind a large desk, listening intently and pressing him for every detail. To Eickhoff's surprise, he seemed more pleased than disappointed. Finally he asked if the experiments had proved with any certainty that the .45-caliber pistol cartridge would function properly with the Blish system, and Eickhoff replied that they had.

"Very well. We shall put aside the rifle for now and instead build a little machine gun. A one-man, hand-held machine gun. A trench broom!"

At once Thompson rose from his chair and sprayed the room with imaginary bullets; he fired from the hip, like a movie gangster of the future.[38]

2

A Broom for Sweeping Trenches

If we had had Thompson guns in 1914, the Germans could never have taken Belgium.
—*Belgian army officer, 1923*

John Thompson had not conceived the idea for a hand-held machine gun quite as spontaneously as it seemed to his chief engineer that September day in 1917. He had followed the war closely, especially on the western front, where both forces were completely bogged down in the mud and barbed wire of trench fighting. The reason for the bloody stalemate, Thompson believed, was the lack of any powerful offensive weapon on either side. Both lines were well dug in against artillery assaults and defended by new Maxim-type machine guns that swept No Man's Land with a blizzard of bullets. In the face of withering machine-gun fire, the traditional cavalry charge was suicidal, the infantry advance slow and costly. But because of the machine gun's size and weight, it was practically immobile, and its enormous fire power could be used only defensively.

The machine gun's potential counterweapons were the tank and the airplane, but neither was yet adequately developed, understood, or employed. The British tanks clanked around the countryside until they foundered in ditches, and the war in the air was often a private one. Allied commanders were quite aware of the new tactical problems created by the machine gun and of the need to somehow utilize its firepower offensively. The light machine gun was one step in this direction. Another was the machine rifle. But these guns still were heavy and clumsy; some were totally unreliable; and none increased the offensive firepower of the individual infantryman as would the miniature machine gun that Thompson had in mind. To Eickhoff he spelled out in detail what he wanted, and why. It would have to be so small and light that an infantryman could go "over the top" firing from the hip, rush an enemy trench behind the shield of his own firepower, and then "sweep it clean" with bullets.[1]

Eickhoff returned to Cleveland immediately to start work on a gun that would represent an entirely new class of military firearm.

The man assigned to the actual designing of the new gun was Oscar V. Payne, a twenty-three-year-old self-taught draftsman and engineer from Centerville, Iowa.[2] At the age of sixteen Payne had taken a job with the American Propeller Company of Washington, D.C., as a general handyman. One day the owner, Spencer Heath, walked into the drafting room unexpectedly and caught his young handyman bent over a drawing table, using drafting instruments to lay out a small machine shop project to which he had been assigned. Payne was promptly promoted to draftsman. A short time later he invented a clever direct-reading pitch indicator for propellers which the company used for many years, nicknaming it the Oscar.

In 1914 Payne went to work for the Knight brothers, Washington patent lawyers specializing in ordnance work. At this time the Knight firm was representing the government in a suit alleging that the Ordnance Department had infringed on firearms patents owned by Benjamin Roberts when it converted the old Springfield muzzle-loader of Civil War vintage into a breech-loader. After Payne had thoroughly studied the patents involved, he was sent to the Ordnance Department to disassemble and make detailed drawings of the only Roberts rifle available. This the Army refused to permit.

"Nothing doing," the officer in charge told Payne's superior over the telephone. "No young kid is going to come in here and mutilate a rare antique firearm." Payne's boss replied that the young kid knew more about guns than the officer and all his so-called ordnance experts put together, and that this was official government business. After some further argument the officer relented.

"Take the gun out of the case and let him work on it," he told the other officers. "But watch him!"

Payne laid the rifle on a table and disassembled it with ease, despite "four or five officers breathing down my neck." One of them he heard whisper, "Look at that, will you? We could never get the damned thing apart."[3]

Apparently the story of this minor feat reached John Thompson, who had just returned to active duty in charge of the Ordnance Department's Small Arms Division. One day late in July 1917, Payne met Thompson while at the Ordnance Department on patent business and mentioned that he was about to leave the Knight firm to enlist in the Army.

"Wait just a minute," Thompson said. "What was your name again?" Then the colonel pulled a card from an indexed file on his desk and quickly scanned it. "Don't enlist. I know all about you, and I have something in mind that would let you use your talents and also serve the country. Go back to your job and I'll call you when I'm ready."[4]

Two weeks later Thompson instructed Payne to get ready to leave town on short notice. After two more weeks, near the end of July 1916, Payne received word to report to Theodore Eickhoff in Cleveland.

Oscar Payne's first job at Auto-Ordnance was to redesign the Blish lock used in the Model I Autorifle which had burst during tests. He did this quickly and successfully, designing a new H-shaped wedge that straddled the bolt and locked it from the front instead of the rear. This new arrangement not only permitted a shorter receiver but linked the locking piece to the bolt so that the two worked in unison to prevent premature detonation of the cartridge.

Before work was finished on the improved Model 2 Autorifle, Eickhoff returned from his trip to Washington with Thompson's instructions to build a minature, pistol-caliber machine gun—immediately. Payne shelved the Autorifle and went straight to the drawing board. Less than two weeks later, on September 22, 1917, he had a tentative design completed and an ink drawing of what he imagined the finished gun would look like. From appearances, it would shoot death rays instead of bullets.

The design of Thompson's "trench broom" was dictated by the novel combat role he expected it to play in trench warfare. It had to be light enough to run with and short enough to fire from the hip, and it had to carry a large supply of ammunition. The gun Payne designed met these requirements nicely. It resembled a large streamlined automatic pistol with a finned barrel, a magazine just forward of the trigger, and handgrips front and rear. It would employ the redesigned Blish lock with the H-shaped wedge and fire .45-caliber Colt automatic pistol cartridges fed from a short belt. Payne soon gave it a suitably dramatic nickname—the Persuader.

Machining of the Persuader began in early October 1917, and within a few weeks the Auto-Ordnance shop had built a working model. Only it didn't work. The gun would chamber, fire, and eject up to seven rounds perfectly, and then jam. A belt feed had been built into the Persuader—partly to eliminate cartridge handling between factory and battlefield, partly to facilitate loading the weapon in the dark, and partly because with machine guns belt feed was simply customary. But the Persuader's internal parts were too light and moved too rapidly to advance the belt evenly, and the reciprocating bolt sooner or later snagged in the fabric.

In December, after three or four weeks of testing, the Persuader was scrapped and work was started on a new model. The intention was still to use a belt feed, but this time the bolt would not enter the belt to push the cartridge out of it and into the chamber. Instead, a small plunger traveling below and just ahead of the bolt would drive each cartridge out of its web pocket and onto a ramp, where the advancing bolt could pick it up as it closed on the chamber.

Before this even more awkward belt-feed system could be installed, Payne, anxious to try out the gun's basic mechanism, improvised a simple spring-loaded box magazine. It worked so well that work on the belt system was postponed and later abandoned altogether. By the middle of spring 1918, the firing and feeding mechanism was showing definite signs of success and taking on the appearance of a finished gun. Auto-Ordnance shop personnel dubbed it, again at Payne's suggestion, the Annihilator.

Both the machining and testing of the first Autorifle mechanisms had been done in the main plant of the Warner & Swasey Company. By the time the "trench broom" idea was conceived, however, Auto-Ordnance had moved most of its operations into the second story of an old frame house owned by Warner & Swasey and used as a woodworking shop. The house was also on Cleveland's Carnegie Avenue, directly across the street from the main plant.

The second-floor walls were covered with patterned wallpaper, faded by the light from the bare windows. In the middle of the main room was a narrow bench supporting the test-firing mechanism, the barrel of which pointed into a steel pipe. The pipe was about six inches in diameter and twelve feet long and passed through an opening in the wall into an adjoining room. There it led to a four-foot-square wooden box filled with sand and reinforced with steel plates. To fire a shot, the firer stepped behind a wooden screen near the test mechanism and pulled a cord connected to the trigger.

Most of the test firing was done by Charles Tunks, the Auto-Ordnance "tinkerer." When Thompson sent Eickhoff to Cleveland on a permanent basis, he told him to hire a good "gun tinkerer" to work the bugs out of new designs by trial and error. Eickhoff located Tunks through Warner & Swasey, and he turned out to be ideal for the job. Patient and painstaking, "he could stand there at a workbench for hours, filing and fiddling until he got something to work."[5] And getting things to work was Eickhoff's biggest headache.

The initial tests of the trial Annihilator convinced Eickhoff and Payne they were finally on the right track, but refining the basic mechanism into a perfected prototype was another matter. Countless minor problems arose as each improvement or modification showed up design flaws in other parts of the gun. But among the employees of Auto-Ordnance excitement was high and rising. "I remember our first disappointments," Eickhoff has said. "Four or five shots—*dadadadadat!* Then nothing. Some part would jam, or it wouldn't feed."[6] But by the summer of 1918 all of the major problems had been overcome. From then on, efforts were concentrated on refining the gun's external features and increasing its durability. "We had no big celebration, but our hearts were high, like when stocks go up. What a pleasure it was to show Thompson that thing actually working!"[7]

John Thompson had stayed in close touch with the Cleveland operation by mail and through Eickhoff's personal visits to Washington. As the machine-gun project progressed, he began making frequent trips to Cleveland himself. Usually he, Payne, and Eickhoff would drive out to the farm of William Burger, the Warner & Swasey engineer, and there test some new feature of the gun, or demonstrate it for someone Thompson had brought along.

In these tests and demonstrations the Annihilator proved worthy of its lethal-sounding name. It had lost the sleek streamlining of Payne's original design, as well as the crudeness of the prototype Persuader, and had taken on the lean and wicked look it would retain in later models. A touch of the trigger unleashed a roar of flame from the muzzle and a shower of empty cartridge cases, emptying a twenty-round magazine in less than a second.

But Thompson, perfectionist to a fault, was never quite satisfied. Time after time he sent Eickhoff and Payne back to the shop to make some modification that would strengthen some part, simplify the gun's disassembly, or increase its reliability. "Time is of the essence," he kept reminding, and "cost no object," although Eickhoff soon learned that any spending Thompson judged to be nonessential would earn him a fatherly lecture on the subject of "O.P.M."—Other People's Money.

By the fall of 1918 the Annihilator handled and functioned to Thompson's satisfaction, and the Auto-Ordnance shop was hard at work on copies of the prototype. In Europe the complexion of the war had changed dramatically since the failure of Ludendorff's great summer offensive, and now the collapse of Germany appeared imminent. Before this happened, Thompson was most anxious to get at least a few handmade samples of the new gun to the American troops, both to combat test its functioning and to measure its effect on the enemy.

After half a century, the surviving members of the original Auto-Ordnance staff no longer can remember the exact sequence of events during the closing weeks of the World War. But they remember well enough their excitement in the fall of 1918, working day and night, and watching the newspapers. On September 30 Bulgaria collapsed; a month later, Turkey. On November 1 Hungary and Austria broke with Germany to sue for peace, and within days Germany itself was torn by bloody uprisings in Kiel and Hamburg. Even if the "trench broom" already was too late to play its intended role in the trenches of northern France, to get it into action at all would mean something to Thompson and to the men who had worked so long and hard on it. Oscar Payne is not certain any more, but believes that one small, secret shipment of several prototype guns reached the New York docks on November 11, 1918, the day the armistice was signed. He does remember clearly the Auto-Ordnance staff's mixed feelings of elation and gloom when the news came in. He turned to Eickhoff and said what everyone there was thinking: "It looks like we missed the boat."[8]

Although, to his great disappointment, his revolutionary hand-held machine gun never made it to the front, Thompson's other work contributed significantly to winning the war. His efforts at Remington not only assured the British an abundance of Enfield rifles, but prepared him for his new job as Director of Arsenals, to which he was appointed by the Chief of Ordnance when recalled to active duty in 1917.

This new post presented Thompson with the responsibility, and the enormous problems, of supplying the Army with all the rifles, pistols, and small-arms ammunition needed to fight a major war for which few preparations had been made. To do this, he launched a controversial production program that immediately drew strong criticism from other military and political quarters. His judgment prevailed, however, and eventually earned him the Distinguished Service Medal.[9]

When Thompson undertook the organizing of the Army's rifle production, existing government arsenal facilities could supply only a small fraction of the 1903 Springfields needed to equip thousands of new combat troops. Private industry would, as usual, have to make up the difference. But the country's large arms manufacturers were tooled up for the .303-caliber British Enfield, and to convert to Springfield rifle production was simply impossible on short notice, not to mention the enormous costs involved in complete retooling. Thompson opposed the obvious solution: that the United States Army adopt the British rifle. He regarded both the Enfield rifle and the rimmed .303 cartridge as inferior and obsolete. He pointed out that the British themselves had planned to abandon the cartridge when the war compelled them to make do with existing ordnance, and that the expedient of adopting a foreign caliber would create enormous problems of ammunition supply.

Supported by Chief of Ordnance William Crozier, Thompson held up production, except on the Springfield, for four months while he supervised the development of an entirely new U.S. rifle, the Model 1917. Incorporating as many of the Enfield's good features as possible, but chambered for the .30-'06, the new rifle was far superior to the British Enfield and could be mass produced in existing factories so quickly and cheaply that it more than made up for the initial delay. During July, August, and September of 1918, the number of rifles manufactured for the United States was 233,562, compared to 112,821 in England and 40,500 in France. By the end of the war the production rate of M1917 rifles had reached 10,000 per day. Moreover, the M1917 cost the U.S. government about $26, compared to $42 paid by the British government for the inferior Enfield.[10]

Thompson was promoted to brigadier general in August 1918 and released from active duty the following December, a few weeks after the armistice. In March 1919 he received the Distinguished Service

Medal

> . . . for exceptionally meritorious and conspicuous service as Chief of the Small Arms Division of the office of the Chief of Ordnance, in which capacity he was charged with the design and production of all small arms and ammunition thereby supplied to the United States Army, which results he achieved with such signal success that serviceable rifles and ample ammunition therefore were at all times available for all troops ready to receive and use them.[11]

General John Thompson returned to civilian life a minor war hero (one feature writer labeled him "Uncle Sam's Premier Gunman") and a businessman with a problem: what to do with a "trench broom" now that the trenches no longer needed sweeping. The money already invested in the gun's development—nearly half a million dollars—demanded some positive thinking as to its peacetime sales possibilities.

For a time after the war, the domestic military market still seemed promising. The country was vowing, as ever, never to be caught unprepared again—to maintain a modern, well-equipped army and navy that would deter any potential aggressors. And for this peacetime duty, it seemed to Thompson, a hand-held machine gun would be ideal. Its portability and firepower would make it an invaluable weapon in border clashes, in putting down insurrections, in controlling riots, or in any situation where mobility and firepower were paramount.

The new gun also seemed to have great potential as a police weapon. After the war the crime rate began to increase alarmingly throughout the country, partly because armed robbers were discovering the convenience of fast motor cars in their line of work. Even when the police arrived at the scene in time to shoot at fleeing bandit cars, their revolvers proved ineffectual against them. What the police needed, obviously, was a little hand-held machine gun. Civil disorders were also on the increase, and certainly a mob of "labor rioters" would think twice before charging a police officer wielding such a formidable weapon. Any time one man or a few men found themselves at a numerical disadvantage, a miniature machine gun would be a very handy item.

In 1919 Thompson put his Auto-Ordnance staff back to work modifying the new gun for use other than in trenches, and put his own mind to work on a marketing problem: what to call such a weapon. Because he objected to calling it a machine gun, for machine guns typically were heavy weapons firing rifle ammunition, he suggested the term "subcaliber machine gun." To this Eickhoff objected, pointing out that "subcaliber" often was used in reference to the adaptors used for firing small ammunition in large guns for target practice. Other terms, including "machine pistol" and "autogun," were considered. The term finally decided upon was "submachine gun," to denote a hand-held, fully automatic small arm chambered for pistol ammunition.[12]

The naming of the submachine gun was accomplished, without much ceremony, at a meeting of the Auto-Ordnance board of directors. The meeting was held in Thomas Fortune Ryan's private New York office, which occupied the top floor of 501 Fifth Avenue. Eickhoff and Payne attended as special guests, and were somewhat awed by their surroundings. Every room in the suite was richly furnished with Continental antiques and Oriental carpets.

Ryan called the meeting to order and quickly disposed of several minor business matters. Then the board turned its attention to giving the gun a name. Thompson already had talked of naming it after the man who had financed it, and Ryan had already rejected the idea. At the meeting he again asked Thompson for a suggestion.

"I propose we name it the Payne Submachine Gun," said Thompson, "because he has gone through all the labor pains of giving birth to our precious baby. No pun intended."

When the laughter subsided, Oscar Payne asked permission to speak.

"What you say is true, General, but you've forgotten one thing. A baby always takes the name of its father." More laughter. "Ours is a new company and it can use a name with prestige. The General's name already is internationally known."

With this, Ryan resolved the matter quickly.

"Mr. Payne, your modesty is gratifying. If that's the way you would have it, that's the way it will be. We'll call it the Thompson Submachine Gun."[13]

Thompson made a halfhearted protest, and then accepted the honor graciously. He would not, however, accept credit for inventing the submachine gun—only for "developing" it. Its inventors, he insisted, were Eickhoff and Payne.

John Thompson's was the world's first "submachine" gun, for he coined the term; but strictly speaking it was not the first hand-held pistol-caliber machine weapon, which the term has come to designate. Nor did this class of weapon make its appearance without precedent or forerunner.[14] Handguns with attachable shoulder stocks were common in the nineteenth century, constituting a weapon somewhere between a pistol and a rifle. The German long-barreled Lugers and Mausers with shoulder stocks and special large-capacity magazines came a step closer to the "machine pistol," which later became the German term for submachine gun. In fact, some Mauser pistols were equipped with a selector switch to permit full-automatic fire.

In 1915 the Italians introduced the Revelli, a strange-looking 9mm affair with twin barrels and spade grips. This was a one-man machine gun chambered for pistol ammunition, but it was ordinarily fired from a fixed mount or a bipod. About the middle of 1918 the Germans introduced the Bergmann Machinen Pistole 18, popularly called the Kugelspritz, or bullet-squirter, built along the lines of a carbine. It was awkwardly designed and crudely constructed, but it fired 9mm Luger ammunition full-automatic and might have proved a formidable trench weapon had it reached the front lines sooner and in greater numbers.

The Germans developed the Bergmann in response to the Italian Revelli and the light machine guns of the French and British. The trench war in France had not only demonstrated the effectiveness of automatic guns but taught both sides that firepower would have to be made portable before it could win any ground. Even the United States Army conceded this, once it was at war. In 1911 an American army officer, Colonel Isaac Newton Lewis, developed the first "light" machine gun which could be carried, set up, and fired by one man, accompanied by an ammunition bearer. But the prototypes worked less than perfectly in early tests, and the Ordnance Department rejected the gun, ignoring its potential. After 1914, however, the Lewis quickly became a favorite with the British, and before the war was over, it was adopted by the United States as well. With much less delay the Army tested and adopted John Browning's gas-actuated machine rifle, the BAR. The Army also gave serious consideration to an adaptor known as the Pedersen device, which converted the '03 Springfield rifle to fire special .30-caliber pistol-type cartridges semiautomatically. The Pedersen conversion would have given the infantryman a low-powered "autorifle," to use John Thompson's term, but it would have required a rifleman to carry two types of ammunition, the conversion mechanism plus special magazines, and also the rifle's regulation bolt for firing .30-'06 cartridges. Knowing the combat soldier's habit of losing things, the Ordnance Department eventually scrapped the idea.

Although the trench war taught the Army to appreciate automatic weapons, none of those either considered or adopted even remotely resembled Thompson's submachine gun. The Pedersen conversion was strictly a semiautomatic rifle, the BAR a heavy machine rifle weighing nearly twenty pounds with bipod and loaded magazine, and the Lewis a light machine gun weighing nearly thirty pounds with bipod and loaded magazine. At the end of the war, only the German Bergmann and the prototype Thompson belonged to the new "submachine" class of small arm that evolved out of the trench fighting on the western front, although both the Italians and the Austrians had developed guns similar to the Bergmann by the time the armistice was signed.

Some firearms historians contend that the submachine gun evolved out of efforts to develop light automatic aircraft weapons, but the relation, if any, is remote. In the summer of 1910, mostly as a stunt, Glenn Curtiss took off from Sheepshead Bay, Long Island, in his new pusher-type biplane carrying Second Lieutenant Jacob Fickel and an '03 Springfield rifle. Fickel amazed spectators by hitting a target on the ground with two shots out of four, but failed to impress the military. In June 1912 Colonel I. N. Lewis prodded the Army into conducting an unofficial test of his new light machine gun fired from a Wright Model B. This was a widely publicized aviation "first," but it again failed to impress military observers. The Army conceded only that it refuted speculation that the sustained firing of a machine gun might cause a plane to tip over in the air.

Early in World War I, once the pilots of observation planes hit upon the idea of shooting at one another, attempts were made to modify pistols for aerial use by equipping them with an oversize magazine and a wire container to catch ejected shells. But once the idea caught on, both sides quickly armed their planes with conventional light and medium machine guns. About the only link between the pistol-caliber submachine gun and aircraft armament was the Revelli, which the Italians developed initially as an air-to-air weapon. Its 9mm pistol ammunition proved too low-powered for aerial use, however, and it was employed thereafter as a kind of light machine gun by the infantry.

No doubt many gun designers had at least pondered the idea of a hand-held machine gun similar to Thompson's "trench broom" or the Bergmann "bullet-squirter." Hiram Maxim himself attempted to build such a weapon prior to World War I. Maxim never publicized his intentions, however, for this historically interesting project did not come to light until 1953, and then only by accident. That year Val Forgett, Jr., now president of Service Armament Company of Ridgefield, New Jersey, attended an auction held by Maxim's estate, near Lake Hopatcong. Among the things being auctioned were many of the personal belongings of the old inventor, including some of his earliest prototype machine guns, which Forgett wanted as a collector. While looking around in Maxim's house, Forgett discovered a most unusual weapon partially hidden under a bed. It was about the size and weight of a Thompson and closely resembled one in profile, but the toggle-type action was that of an over-size Luger pistol in almost every respect. The frame was aluminum and the bore .22 caliber. It was chambered, however, for a centerfire cartridge which Forgett guesses to be the 5.5mm (.22 caliber) "Velo Dog," a light round marketed in Europe about 1900 for the express purpose of arming cyclists against hostile canines. The gun was obviously homemade and only about 80 per cent complete, lacking a foregrip, a finished feed mechanism, and probably other parts. It apparently never had been fired, and there is some doubt that the gun's low-powered cartridge could have operated the rather heavy breech mechanism. Nevertheless, it was clearly an effort to build a submachine-type weapon. The project may have been abandoned due to design problems, or due to Maxim's death in 1916.

According to George Goll and Theodore Eickhoff, Thompson had no personal association with Hiram Maxim that might have led either man to inspire the other toward weapons quite similar in type and appearance. The Maxim gun's Luger action is also something of a mystery. Most arms historians have considered the Borchardt and Luger designs of the eighteen-nineties to have been influenced by Maxim. If the experimental Maxim gun could be dated earlier than Luger's and Borchardt's pistols, the Luger especially would be revealed as a close plagiarism. Otherwise it appears that Maxim was not averse to copying designs himself. Another mystery is why Maxim chose to chamber the gun, with its heavy one-inch-diameter barrel, for a low-powered and rather uncommon .22 centerfire cartridge.[15]

If the World War inspired the submachine type of weapon, the armistice retarded its further development in every country except the United States. The Allied arms-control agency destroyed the Austrian prototypes and the tools for their production at the Steyr works, and the treaty of Versailles virtually prohibited Germany from building or buying machine weapons. In Italy, the Revelli became extinct, and its later hand-held version (a 1918 modification by the Beretta arms factory) was never produced on a large scale. Which left the American Thompson, for all practical purposes, the only submachine gun still under development or having any prospects for the immediate postwar future.

After the armistice, the Auto-Ordnance staff in Cleveland, now working at a more relaxed pace, continued refining the submachine gun into a weapon more versatile than the simple "trench broom" originally planned. Throughout 1919 and 1920 the gun underwent numerous minor alterations. Three separate groups of prototypes were produced. These were listed in Auto-Ordnance records as Annihilators I, II, and III.[16] According to an interoffice report dated

November 12, 1919, twenty-eight prototypes had been scheduled for construction, and most of these already had been built. Apparently all were marked "Model of 1919," regardless of their variations.

Annihilator I was the finished version of the first workable test mechanism and was marked with the serial number 1. It preserved some features of the original Persuader, including the perforated-jacket cooling system and a provision for belt feeding, which Auto-Ordnance then still hoped to perfect eventually. The Annihilator II group consisted of two guns, serial numbers 2 and 3. In these, all belt-feeding parts were eliminated, and the cooling jacket was replaced by radiation fins milled into the rear half of the barrel. Except for their hook-shaped cocking lever and the absence of sights and buttstock receptacle, submachine guns #2 and #3 were in appearance virtually identical to Thompson guns of the future. Gun #2 was used as a demonstrator, and #3 for various experiments with silencers, bayonet attachments, improved firing pins, and so forth.

The improvements suggested by extensive testing of gun #3 were incorporated in the building of ten more prototypes, designated Annihilators III, and in April 1919 work began on fifteen more guns of the same design, which were designated Submachine Guns D. From the Annihilator III group, two guns, serial numbers 10 and 11, were modified in firing pin and actuator and designated Submachine Guns E. These two guns apparently became the prototypes for the actual production model later.

Two other Annihilator III guns were modified to become a separate subgroup and given the designation Submachine Guns F, for Freak. These were an idea of Oscar Payne, who believed he could put together a functional submachine gun with as few as eleven parts. During one of John Thompson's visits to Cleveland in the spring of 1919 he and Payne discussed the simplicity of their submachine gun compared with other automatic weapons. Payne remarked that the gun probably could be simplified even further were it not for certain Ordnance Department design requirements. Thompson mulled this over during his train ride back to the company's New York office. When the train stopped at Buffalo, Thompson sent Payne a telegram instructing him to build an experimental gun with as few parts as possible for demonstration and advertising purposes.[17] Within a few weeks Payne had two simplified submachine guns ready for testing, and both worked well. Gun #8 consisted of only eleven parts and gun #9 of seventeen, compared with about thirty in the other prototypes. Neither gun went beyond the experimental stage, although the eleven-part model was widely publicized in early Auto-Ordnance promotional literature.

One important step in the evolution of the prototype Thompson was the addition of a novel self-oiling feature that automatically lubricated the gun's moving parts. To John Thompson, who liked to draw analogies (and sometimes stretch them), a firearm was a gas engine not unlike that of an automobile. Both engines utilized a closed cylinder in which combustion took place, the expanding gases driving a piston in one case, a bullet in the other. Having discovered this much similarity, Thompson began to wonder why the gasoline engine of an automobile generally worked for long periods without malfunction while the gas engine of the machine gun was so susceptible to stoppage. One reason for this, he decided, was that machine guns had no built-in system of lubrication.[18] He put the Auto-Ordnance staff to work on the problem, and Oscar Payne soon devised such a system in the form of two oil-soaked felt pads mounted inside the submachine gun's receiver. Each time a shot caused the bolt to move back and forth, it brushed against the felt pads and squeezed a tiny amount of oil into the gun's mechanism. Whether or not Thompson's gas-engine analogy was sound, the self-oiling feature made the bolt and firing mechanism exceptionally smooth in operation and contributed a good deal to the submachine gun's reliability.

An even more important and more conspicuous Payne invention was the submachine gun's revolutionary drum magazine. In existing drum magazines designed for vertical attachment, the spring pressure usually was applied to the last cartridge, which tended to force the other cartridges outward against the wall of the drum. This presented a friction problem and caused the cartridges to be delivered to the magazine outlet with increasing force as their number diminished. Payne overcame this difficulty by employing a six-armed rotor powered by a clock-type spring. The cartridges, loaded in independent groups of five between the arms of the rotor, traveled in an outwardly spiraling track to an opening at the top of the drum to be picked off one by one by the advancing bolt. The rotor's arms thus distributed pressure equally to cartridges in all parts of the track, delivering them at a uniform rate. And because the drum contained its own source of power, the clock-type spring, it required no connection to the gun's bolt mechanism and could be used interchangeably with a box magazine.

Payne designed the drum magazine in two sizes, to hold fifty and one hundred cartridges. These were designated Type L and Type C, respectively, conforming to the Type XX designation given the earlier twenty-round box magazine. When the submachine gun's drum magazines were unveiled in trade and military journals, firearms experts were especially impressed by the Type C. "The raising of 100 cartridges against gravity," reported the *Army and Navy Journal,* "is a remarkable and hitherto unaccomplished feat."[19] *Army Ordnance* called the Payne magazine "a triumph in its own field."[20]

Very little of the submachine gun's development took place at Warner & Swasey. By the time work started on the Persuader, in October 1917, Eickhoff already had acquired a few machine tools and established a small shop, and by the spring of 1919 Auto-Ordnance was largely self-sufficient, with a staff of about twenty-five. The company still maintained a testing facility in the old frame house across from the Warner & Swasey plant on Carnegie Avenue, but its offices had been moved from the Warner & Swasey building to the Meriam Building on Euclid Avenue, and a more complete shop had been set up at the Sabin Machine Company, farther east on Carnegie.[21]

Theodore Eickhoff was chief engineer, and Oscar Payne his chief assistant. Working directly under Payne on the submachine gun were three draftsmen, Elmer Koenig, Corless Hancock, and Doren Curtis. Fred Deertz, assisted by William Dute, supervised the machine shop at the Sabin Company and a staff of ten to fifteen machinists. Charles Tunks tinkered and gunsmithed, and George Goll acted as liaison between Thompson and Auto-Ordnance staff and did a little of everything.

Eickhoff remembers the Auto-Ordnance staff as group of likable, energetic young men who considered their work an interesting challenge. They got along well with one another and functioned efficiently as a team, except for Oscar Payne, who was something less than an organization man. Versatile as he was talented, Payne did not always feel the need of supervision, and in fact tended to supervise everyone else. To discourage this practice as tactfully as possible, for Payne's ability was invaluable, Eickhoff would put another man to work independently on the same project as Payne, and then not tell Payne how well his competition was doing.

Early in 1919 Eickhoff put Payne back to work on the Thompson Autorifle. Auto-Ordnance had shelved the project in the fall of 1917 to concentrate all efforts on the submachine gun, and at that time the rifle's prospects did not seem at all promising. Payne had solved the problem of premature detonation, which had caused the first prototype to explode on the testing stand, but there remained the apparently insurmountable problems of reliability and rifle cartridge lubrication. Privately Eickhoff considered the rifle a total failure, but John Thompson still had complete faith in the Blish system. As soon as the war was over, he ordered work resumed on the rifle and designing started on a complete line of Blish-lock firearms, including a two-pounder "Autocannon," a .50-caliber heavy machine gun, and a semiautomatic pistol, shotgun, and .22-caliber rifle.[22]

During 1919 the submachine gun still had top priority, but the Autorifle took precedence over Thompson's other proposed firearms. The Model I Autorifle, redesignated the Model A, was abandoned. The Model B prototype, incorporating Payne's integrated bolt and lock, was dusted off and completed. It functioned more or less satisfactorily in tests, but weighed an unacceptable sixteen pounds. A lighter, simpler version, the Model C, was built next, but it only proved beyond any doubt that the sliding wedge-type Blish lock was not practicable in a high-powered rifle.

There remained, however, two rotary versions of the Blish lock still to be tried. In one, proposed for a Model D prototype, a wedge with inclined locking lugs at the rear of the bolt rotated into an unlocked position; in the other, wedge and bolt were combined in a rotating bolt with interrupted threads. This latter design operated on the same principle as the interrupted-thread-type breech blocks on the large naval guns that years earlier had inspired John Blish by unscrewing themselves under reduced pressure. In extensive tests this rotating-bolt system worked well, and by the end of 1919 the first properly functioning prototype of the Thompson Autorifle was completed. It was designated the Model P, for Oscar Payne.

By this time also the .22-caliber semiautomatic rifle existed in prototype. It was a sleek and sporty 5.5-pound target gun inspired by John Thompson's belief that the ending of the war would bring an upsurge in sport shooting and target practice. It functioned reliably except for recurring ejection problems, probably caused by its miniature rotating Blish lock. Payne had argued with Thompson that with such a low-powered cartridge a lock of any kind was superfluous, but Thompson would not have a rifle without one.

None of the other proposed guns reached even the prototype stage, although some Autocannon experiments were conducted. In 1919 Auto-Ordnance bought an old two-pounder artillery piece from the Bannerman arms firm in New York, equipped it with a crude form of Blish lock, and fired it a few times on the William Burger farm outside of Cleveland. The tests received the usual optimistic mention in a report to Thompson, but nothing further came of the gun.

After the war and his release from active duty, Thompson, early in 1919, moved from his country home in Pennsylvania to a modest but attractive two-story frame house on Bank Street in New Canaan, Connecticut. He also maintained a town house on Manhattan's upper East Side, near the Metropolitan Museum. His military duties over, he at last had time to pursue his personal interests, and to participate more fully in the affairs of Auto-Ordnance.

In New York, with his son, Marcellus, and Thomas Fortune Ryan, he concerned himself with matters of management—manufacturing contracts, promotion, marketing, sales. In Cleveland, he acted as consulting engineer and gun tester, packing Payne, Eickhoff, and Goll into a car loaded with guns and equipment for lengthy afternoon shooting sessions at the Burger farm. And at his home in New Canaan, he became, to both the amusement and dismay of everyone else, an energetic backyard inventor.[23]

As a developer of small arms and equipment, Thompson had earned himself a worldwide reputation for developing excellent weapons and expertly managing their manufacture, but he was not, in the usual sense, an inventor. His genius lay primarily in conceiving a new design or process and then supervising some talented young draftsman-engineer in working out the details. At Auto-Ordnance he would quickly abandon his efforts to make a sketch and hand the pencil to Oscar Payne, then lean back in his chair and tell him what to draw. Thus Thompson developed, Payne invented, and Eickhoff saw to it that the image became an object.

But Thompson was an inventor at heart—a most original and creative inventor—and now, with free time on his hands, he saw no reason to hold back any longer. From New Canaan came rough sketches for one new invention after another—most of them exhibiting the imagination of Da Vinci and the practicality of Rube Goldberg. Patiently, Theodore Eickhoff turned the sketches over to his staff for drafting, and patiently the company's patent attorneys composed their applications for patents, a number of which were granted, to Eickhoff's considerable surprise.

One of Thompson's inventions was a self-consuming rifle cartridge. Brilliantly conceived, its development awaited only the discovery of a slow-burning, stable, plastic explosive material and a means of detonating it. Other inventions included a semiautomatic double-barreled shotgun, a semi-dirigible airplane, and a hollow, tube-shaped airplane fuselage with an air duct from nose to tail. The invention Eickhoff best remembers, with a smile and a sigh, was Thompson's ingenious gun-whistle, patent 1,334,059. This was a special firearm or firearm attachment which utilized the barrel gases to blow a whistle. Its purpose was to signal or summon aid in the unlikely event that the gunshots failed to attract anyone's attention.

Fortunately John Thompson never permitted his inventive impulses to sway his technical or business judgment, which is probably why none of his home inventions got any further than the drawing board.

After 1920 Thompson was joined by his son, Marcellus. The younger Thompson, disenchanted with the Army, had resigned his commission in December 1919 to become vice president and general manager of Auto-Ordnance. He undertook the job with great energy and enthusiasm, but lacked his father's warm personality and executive ability. As a West Point cadet nicknamed Tompo, he "boned the goat position" throughout his four years and graduated next to the bottom of his class in 1906. According to a classmate, Marcellus nevertheless considered himself the rightful winner of last-place prize. The only cadet below him, he would point out, had been turned back from the class of 1905.[24]

Unlike his father, Marcellus was cool and aloof in his relationships with others, and had invited a good deal of harassment during his first year at West Point by refusing to exhibit the humility expected of a plebe. He possessed a sense of humor, but revealed it only in the company of his few close friends. Others found him to be reserved and taciturn, a man of great determination verging on simple stubbornness. Throughout his life he was handicapped by having such a prominent and popular father, to whom he was frequently, and usually unfavorably, compared. Theodore Eickhoff, who loved John Thompson as his own "second father," considered Marcellus a man of "big ideas and a big mouth."[25] Others remember him more kindly: "He was not an easy man to know, but his close friends truly loved and respected him."[26]

As an artillery officer in the peacetime Army, Marcellus Thompson found his first ten years of military life singularly unexciting. After receiving his commission he served a year in Cuba, then returned to the States to a series of undesirable coast artillery posts in the Florida Keys. In 1911 he was transferred to Fort Washington and soon became a prominent member of Washington, D.C.'s cocktail and drawing-room set. There he met and courted Dorothy Harvey, whom he married on August 12, 1914. The wedding was a major event in eastern social circles. The elaborate military ceremony took place at the country estate of the bride's father, Colonel George Brinton McClellan Harvey, near Deal, New Jersey, and was attended by more than four hundred relatives and guests brought in from Washington and other cities by private trains. *Vogue, Town & Country*, and nearly every metropolitan newspaper in the East published pictures and accounts of the ceremony.[27]

In 1915 Marcellus Thompson was transferred to Fort Hancock, New Jersey, where he won the Army's William Knox trophy for accuracy with twelve-inch guns. In September 1917 he went to England with the Coast Artillery's Seventh Provisional Regiment, then to the training fields at Mailly-le-Camp, France. From October through the following May he served as secretary for the antiaircraft and trench mortar schools at Langres; then he attended the heavy artillery school at Mailly. In the summer of 1918 Thompson took command of a battalion of the Sixty-fifth Artillery and was engaged in several major battles.

Thompson's artillery record earned him promotions to the rank of lieutenant colonel and appointment to the post of Assistant Chief of Staff of the First Army Artillery. After the armistice he was promoted to Chief of Staff. Despite his unspectacular performance at West Point, his war record was excellent, and he outranked most of his graduating classmates when he returned to the United States in April 1919.

After his experiences in France, Thompson's postwar duties seemed less engaging than ever. He became further discouraged when the Army reduced him to his peacetime rank of captain, which successively cost him his commands of Fort Warren and Fort Strong in Massachusetts upon the arrival of superior officers. Within a few months he dropped from commanding officer to post quartermaster, to post athletic officer, and finally to recruiting officer. On December 15, 1919, he resigned his commission and went home to sell Tommyguns.

Marcellus Thompson joined Auto-Ordnance just as the company was taking the first steps toward getting the submachine gun into production. While John Thompson supervised the work on the finishing touches it still needed and shopped for contract manufacturers, Marcellus began exploring its market possibilities and mapping out a promotional campaign.

The most desirable market was, of course, the United States military. Despite the Ordnance Department's long-standing prejudice against automatic weapons, it had adopted the Lewis gun and the Browning automatic rifle and machine gun; moreover, it had expressed considerable interest when John Thompson first proposed his novel "trench broom" around the end of 1917—and Congress was still pledging military preparedness. The Navy, too, was a promising market. The submachine gun would be particularly well-suited to Navy and Marine landing parties, and the Navy already had shown more open-mindedness than the Army toward automatics by adopting the Colt-Browning Model 1895 machine gun while the Army still stuck to its Gatlings.

Also encouraging to Auto-Ordnance was the military response to a series of tests and demonstrations of the submachine gun that began in the spring of 1920. On April 27, 1920, the Ordnance Department conducted a functioning test of the gun at the Springfield Armory and reported only one stoppage in the firing of 2,000 rounds. The report of the test stated that "although the gun was fired as rapidly as magazines could be loaded for each 1,000 rounds, the gun did not heat sufficiently to cause trouble, and . . . the action of the gun appeared strong enough to stand continuous pounding, and functioned reliably during the test."[28]

In the summer of 1920 the Marines tested the submachine gun at Quantico, Virginia, and were equally impressed. Major P. H. Torrey later gave the following account of the gun's field demonstration in *The Marine Corps Gazette*.

> During the writer's observation of the demonstration, some three thousand rounds were fired automatically in bursts of twenty shots, and singly, at ranges from 100 to 400 yards. During the entire demonstration not a malfunction occurred. The gun functioned perfectly at any angle of elevation or depression, canted right, left, or upside down. No little difficulty was experienced in catching the exact time due to the rapidity of fire, which varied from 4/5 to 1 1/5 seconds per string. This appears to bear out the claim of the inventors that the gun is capable of firing 1,500 shots per minute. At 100 yards, firing single shots at a "B" target, in the prone position, with stock attached, no trouble was experienced in making perfect scores. At this range, using

automatic fire, one could without difficulty group five shots on a 6′ x 6′ target, after becoming somewhat familiar with the action. At 200 and 300 yards remarkable scores were made from the prone, kneeling and sitting positions, semi-automatically, at the rate of ninety shots per minute. Firing automatically the dispersion increased considerably. What appeared to be the first few shots hit the target and occasionally others. All the shots appeared however, to be striking in the vicinity of the target as the splashes at this distance were clearly visible. The demonstration was concluded with firing at 400 yards, by an experienced shot, slow fire, prone. Some difficulty was experienced at first in getting on the target. After this was accomplished several strings of ten shots were fired which resulted in very good groups. When one considers the remaining velocity of a .45 calibre bullet at this range; that the gun used was fitted with a very crude rear sight, and that there was a tricky fish-tail wind blowing from six o'clock during the entire time, the performance might well be considered remarkable.[29]

The submachine gun's first public demonstration took place at the National Matches held at Camp Perry, Ohio, in August 1920. With George Goll doing the shooting, the gun performed at its best and attracted crowds of onlookers. On the 200-yard range Goll made a perfect score with twenty semiautomatic shots fired in ten seconds; on the 500-yard range he emptied six hundred-round drums in only seventy seconds and still scored a respectable hundred hits. When fired full-automatic, at a cyclic rate of about 1,500 rounds per minute, the gun sounded like "the loud ripping of a rag."[30] Everyone who saw the gun in action marveled at its size and its firepower and agreed that it was the most revolutionary small arm of its day.

Heartened by the submachine gun's enthusiastic reception, John Thompson, about the middle of 1920, approached the Colt company in Hartford with a proposal to manufacture it under contract. The president of Colt's, Colonel Skinner, was a good friend of Thompson, and the Colt imprint was the highest recommendation any firearm could have. Also, Colt's had a closer working relationship with the armed services than did any other small-arms company. But Colt's did not especially want to manufacture the gun under contract. After thoroughly examining and testing it, Colonel Skinner made Thompson a counterproposal: the Colt company would buy all rights to the submachine gun for an even $1,000,000. This would have meant a tidy profit of about $500,000 for Auto-Ordnance, but Thomas Fortune Ryan preferred to gamble. He told Thompson, "If it's worth a million to them, it's worth more than a million to us."[31]

Its offer turned down, the Colt company contracted to manufacture 15,000 basic firing mechanisms for $680,705.85 (about $45 each), plus $9,105 for spare parts. Auto-Ordnance would furnish all tools, fixtures, and gauges at an additional cost of $210,323.93. For the manufacture of sights and walnut stocks, contracts were signed with the Lyman Gun Sight Corporation for $69,063 and with Remington Arms for $65,456. Funds were supplied by Thomas Fortune Ryan and secured by a chattel mortgage on all Auto-Ordnance corporate property.[32]

The contracts were negotiated in the fall of 1920, to everyone's mutual satisfaction. The only disapproval came from one outsider. John Browning, the salty, irritable, banjo-playing genius of automatic firearms, regarded the Colt company as his private domain and John Thompson as an interloper; he made it clear to Colt's management that when he visited Hartford, which was frequently, he did not want to meet or even see anyone from Auto-Ordnance.[33]

After signing the contract with Colt's, Auto-Ordnance closed down its Cleveland operation, released most of its engineering and machine-shop staff, and moved into a rented building on the Colt company grounds in Hartford. It also opened a New York City sales office, at 302 Broadway, and launched a promotional campaign aimed primarily at police agencies and the domestic and foreign military. The campaign advertised the submachine gun primarily, but also announced that Auto-Ordnance soon would market its .30-caliber Thompson Autorifle and .22-caliber semiautomatic sporting rifle, and later its revolutionary "Autocannon," a model of which was said to already exist. The submachine gun was offered in calibers .45, .38, .32, and .22, although no steps had been taken toward manufacturing the gun in any caliber but .45. The .22-caliber model, though never built, even in prototype, was described as measuring sixteen inches without buttstock and weighing only three pounds.[34]

The gun put into production by Colt's was designated the Thompson Submachine Gun Model of 1921. It was a handsome commercial weapon with the carefully blued finish and graceful walnut grips of a sporting arm. Unlike the early prototypes, it had an attachable buttstock, a selector switch to permit aimed semiautomatic fire, and a slotted actuator (cocking) knob centered on top of the receiver. It had a simple blade-type front sight and a folding rear sight adjustable for windage and ranges up to 600 yards—probably the most useless feature ever put on a submachine gun. It fired at the rate of about 800 rounds per minute, as compared to 1,000 and 1,500 r.p.m.[35] claimed for the prototypes. In range tests, the higher rates of fire had proved unmanageable, especially with untrained gunners. List price for the Model 1921, with buttstock and one Type XX magazine, was $225.

The first small batch of finished guns came off the Colt assembly line in the last days of March 1921, and were rushed to Auto-Ordnance salesmen and to the Army and Marines for testing.[36] At Quantico, Virginia, on April 4 and 6, the Marines tried out the new gun both in the air and on the ground. In the first test, a gunner, leaning out of the cockpit of an airplane, fired 1,200 rounds at a land target from an altitude of 300 feet, and then 200 rounds at a sea target from an altitude of 700 feet. Apparently the results were less than spectacular, as the gun was praised only for its lack of vibration. It performed more impressively in the ground test on April 6, when one gunner made a perfect score with 20 rounds at 75 yards. Members of the testing board reported that the gun had possibilities as protection for artillery, as an auxiliary gun in machine-gun nests, and as a weapon for night raiding parties and jungle patrols. On April 8, at Camp Benning, Georgia, members of the Army's Infantry Board tested the Thompson for both defensive and attack fire. In both tests submachine gunners equaled or surpassed the scores of squads armed with rifles and BARs. According to the *Army and Navy Journal*, the board considered the test "very successful" and the Thompson gun "as near mechanical perfection as it was possible to make an arm of its type."[37]

During the month of April 1921 Auto-Ordnance delivered a total of 107 submachine guns. Some of these went to salesmen and military authorities in England, Panama, Costa Rica, Honduras, Colombia, Bolivia, and San Salvador. Most went to large steel and mining companies in the South, where recurring labor troubles had given rise to heavily armed industrial police forces hired to "protect private property" from the depredations of strikers. But in that first month the most consequential delivery was that of seventeen guns purchased by George Gordon Rorke, a young Irishman from Washington who had placed advance orders for five hundred submachine guns and who for months had been hounding Auto-Ordnance to rush him his merchandise.[38]

3

The Irish Sword

See what the Irish crowd think of the gun.
—*Marcellus Thompson, at an Auto-Ordnance sales meeting in December 1920*

I do not wish to discuss the *East Side* matter except to say that we do not know how the guns got aboard the ship.
—*Marcellus Thompson, to reporters in June 1921*

Less than three months after it went to market, the Thompson submachine gun received more publicity than Auto-Ordnance ever could have hoped for—or desired. On June 15, 1921, some 495 of the guns sold to the impatient George Gordon Rorke unexpectedly turned up at Hoboken, New Jersey, in the hold of a Dublin-bound ship, on their way to the Irish Republican Army. The *New York Times* described the smuggling plot as "one of the largest in filibustering history."[1] It was also one of the most mysterious, and potentially embarrassing to the United States and to the prominent persons who found themselves implicated.

Nor could the timing have been much worse. Throughout the Sinn Fein rebellion, Irish-Americans in the United States had been supplying the rebels with guns and money and flooding this country with anti-British propaganda. Several prominent political figures, both state and national, were demanding that the United States recognize the Irish Republic, some even calling for American military intervention. By June, the issue was explosive. The British, anxious to end the unpopular conflict, had stepped up military pressure in hopes of driving the rebels to the conference table, and British politicians were angrily blaming the American Irish for perpetuating the struggle. To aggravate the situation even further, U.S. Rear Admiral William Sims, in a speech in London on June 7, called American Sinn Feiners "traitors" and "jackasses."[2] This pleased the British enormously, but sent the Irish-Americans into hysterics.

Both the American and the British government had been working hard to weather the Irish storm in peace. Washington appreciated Britain's predicament: how to salvage its interests in Protestant, anti-Republican Ulster short of total war against southern Ireland—which English popular feeling would not support, and which would jeopardize relations with the United States. For their part, the British understood American politics well enough to disregard the anti-British railings of politicians. This was the traditional means of getting the "Irish" vote. Also, the British believed Washington's promise that the United States would remain strictly neutral and do all it could to stop the smuggling. But from I.R.A. documents captured in Dublin in May, the British knew the rebels hoped to stage a spectacular fall offensive that would improve their bargaining position. And they knew that the rebels, who were desperately short of arms and ammunition, were counting heavily on a large shipment of Thompson submachine guns coming in soon from the United States.[3]

About June 3 the British government brought this to the attention of Auto-Ordnance through diplomatic channels. On June 6 Marcellus Thompson wrote to the British Consul General to assure him that such a shipment was impossible; that the company had just marketed the gun and had not yet received any sizable orders.[4] He neglected to mention that two weeks earlier the company had delivered fifty guns to three men—named Murphy, O'Brien, and Gallagher, in care of an Irish saloon on Manhattan's lower West Side—and that delivery was almost complete on the five hundred guns ordered by George Gordon Rorke.

When Rorke's guns made their sensational appearance at Hoboken nine days later, Marcellus Thompson had nothing to say to the press. Nor did Colonel George Brinton McClellan Harvey, who had just taken his new post as American ambassador to Great Britain. It could not be concealed that he was Marcellus Thompson's father-in-law, but apparently neither the State Department nor the British discovered that he was also one of the founders and stockholders of the Auto-Ordnance Corporation—and as such, at least technically, a seller of guns to Irish rebels.

In the 1930s, a story circulated in the Auto-Ordnance offices to the effect that Thomas Fortune Ryan himself had been one of the principals in the Great Irish Smuggling Plot; that he and several other supporters of the Cause had managed to get one shipment of Thompson guns safely to Ireland, but their second shipment had been captured by the Coast Guard.[5] Embellished with the customary sea chase and gunfire, the story was told as a bit of interesting company lore, and may have been little more than coffee-break fiction. But it was told by an Auto-Ordnance official who was a member of the Ryan family, and Ryan's Irish Republican sympathies, as well as his friendship with Republican leader Eamon de Valera, were widely known. In addition, Ryan supposedly was one of several prominent American businessmen who secretly contributed large sums of money to the Republican movement.[6] Whether or not the Ryan stories had any truth to them, evidence developed by the Department of Justice in its investigation of the smuggling plot did suggest that the Auto-Ordnance Corporation had a sales department well staffed with "Sinn Feiners."

G. Owen Fisher, a middle-aged, $10,000-a-year sales representative, told government investigators that at one of the company's first sales meetings Marcellus Thompson had instructed him to "see what the Irish crowd think of the gun."[7] This was in December 1920. About the middle of February 1921 Fisher took an order for fifty guns from the three New York Irishmen, John J. Murphy, John O'Brien, and John Gallagher. The guns were delivered in care of saloonkeeper P. J. Gentry on May 25, and apparently reached Ireland.

A month earlier, in April, Fisher had been approached by a man from Chicago named James J. Dineen, who expressed interest in buying a substantial number of guns on behalf of another party. Fisher remarked that Dineen's name sounded Irish, and was told, in so many words, "Perhaps that is why the party who wants the guns has employed me."[8] Obediently, Fisher informed Auto-Ordnance of his prospective customer, and the company arranged for Dineen to be given several training sessions with the gun at New York's Seventy-first Regiment Armory firing range.

Dineen took a pair of guns back to Chicago with him and demonstrated them for two I.R.A. agents, Larry de Lacy and Sean Nunan, the latter of whom would one day serve as Irish ambassador to the United States. Both de Lacy and Nunan fired the Thompson; they decided it was an ideal weapon for the I.R.A., and sent Dineen to Ireland to demonstrate it for the rebels' top political and military leaders.[9]

Dineen, accompanied by another Chicagoan, Patrick Cronin, reached Dublin about the middle of May, the two submachine guns hidden in their luggage. Dineen and Cronin were both former United States Army officers who had become active in the Irish Republican movement in Chicago after the war, and both had reputations as expert gunners. On May 24, in the basement of an isolated house in the suburbs of Dublin, they gave their first lecture and demonstration. Present were Michael Collins, Tom Barry, Richard Mulcahy, Geroid O'Sullivan, and six members of Collins' murderous "Dublin Squad." After stripping, reassembling, and explaining the gun, Cronin attached a loaded magazine and invited Collins to try it out. Collins looked at the new-fangled weapon with uncertainty and replied, "No bloody fear, Tom will do it!"[10] Tom Barry, commander of the Cork No. 3 Brigade, reluctantly took the gun and succeeded in hitting a target of stacked bricks at one end of the basement. In his *Guerilla Days in Ireland*, Barry writes:

> The smashing of the targets by the first shots from a Thomson [*sic*] gun in Ireland was taken as a good omen by all who were present, but my interest in the efficiency of the gun was far less than my concern

> not to miss in front of the party who would rag me unmercifully and probably offer to teach me to shoot. Before we left the building Collins and Mulcahy had decided to purchase five hundred of the Thomsons.[11]

Irish Republicans in the United States apparently had never doubted that the I.R.A. would want Thompsons, for sizable orders already had been placed. John J. Murphy had ordered fifty in February with an option to buy another hundred and fifty. George Gordon Rorke and an associate, Frank B. Ochsenreiter, had ordered a hundred in January and four hundred in April, with an option to buy as many as a thousand. Murphy's guns were delivered to P. J. Gentry for resale to "miners and sportsmen in Alaska." Rorke's and Ochsenreiter's went to a New York "importer-exporter" known as Frank Williams.

Rorke, a handsome twenty-nine-year-old Washington socialite, was a former president of the Protestant Friends of Irish Freedom. He was also a member of the Association for the Recognition of the Irish Republic, one of whose organizers was "importer-exporter" Frank Williams. Ochsenreiter was a member of the same organization and an old friend of Rorke from their law-school days at Georgetown University. He was the manager of the Washington office of Royal Typewriter, a Ryan company, and was acquainted with Ryan personally. He also knew Frank J. Merkling, Ryan's long-time personal secretary, who later became Auto-Ordnance secretary and a member of the board of directors. His cousin, George T. Wise, was the company's Washington sales representative.

These personal and business relationships were uncovered by Department of Justice investigators who, when the smuggling attempt came to light, immediately seized Auto-Ordnance records and picked up several persons for questioning. The case was a particularly delicate one on two counts: the several prominent persons connected with Auto-Ordnance, and the diplomatic question of American military aid to forces in rebellion against a friendly constitutional government. Had the smuggling attempt succeeded, the *New York Times* advised, it would have further strained British-American relations, and perhaps resulted in a reverse version of the *Alabama* case, in which an international court held Britain responsible for $15,500,000 in damage to U.S. shipping done by Confederate raiders operating out of British ports during the Civil War.[12]

From company records and from questioning, Justice Department investigators surmised that Frank Merkling was the main link between Auto-Ordnance and its Irish customers. Merkling pleaded innocent and ignorant. He explained that the Auto-Ordnance sales board had indeed had some misgivings about selling machine guns to some young Irishmen with a lot of cash and no desire to reveal their guns' destination. At an executive meeting with George Rorke present, George Smith, a director of both Auto-Ordnance and Royal Typewriter, noted Rorke's evasiveness and asked him flatly, "How do we know they are not going to Ireland?" Rorke pointed to a Masonic pin in his lapel and quipped, "I am not that kind of an Irishman."[13]

The board, not anxious to lose such a large cash sale, ultimately decided that problems of this sort should be left to the discretion of Auto-Ordnance's sales manager, Walter Morgan. And Morgan, on the advice of Merkling, contented himself with a signed statement from Rorke that the guns would not be resold to Russia, Mexico, Germany, Japan, or to "radicals" in the United States. Morgan and Merkling also decided between themselves to not bring the matter up again.

The first entry in the Auto-Ordnance Corporation's shipping records for 1921 lists the delivery, on March 24, of two guns to the company's Washington representative, George Wise. The address, however, was that of Wise's cousin, Frank Ochsenreiter, who presumably turned the guns over to Rorke. The filling of Rorke's order began on April 2 and proceeded as rapidly as finished guns came off the Colt assembly line, with a few held out for police departments, the armed services, foreign governments, and industrial firms. By June 11, 490 submachine guns, plus a large number of extra magazines and spare parts, had been paid for by Rorke and delivered to the warehouse of Frank Williams at 2088 LaFontaine Avenue in the Bronx.

Frank Williams had many names, but his real one was Laurence de Lacy—the same Larry de Lacy for whom Dineen and Cronin had demonstrated the Thompson gun in Chicago. Formerly the editor of a militant Republican newspaper in Ireland, de Lacy had got himself involved in a rebel dynamite plot about 1915 and had fled to the United States to escape arrest by the British.[14] In California he had edited *The Leader*, a Catholic weekly that tended to be pro-German, and later spent eighteen months in prison for his part in a plot to free two German diplomats, Franz Ropp and E. H. von Schack, from the custody of military authorities. Released in 1919, he moved to the east coast and became active in Irish Republican societies, including the secret Clan na Gael. The Clan had been helping supply the I.R.A. with smuggled arms and ammunition, obtaining funds from wealthy Irish-American sympathizers and through its respectable front organization, the Friends of Irish Freedom (Victor Herbert, president). By 1921, however, the Clan had splintered into several squabbling factions. The most militant of these groups, and the one with which de Lacy worked most closely, had reorganized as the Association for the Recognition of the Irish Republic, and may have been the source of de Lacy's $150,000 cash which Rorke delivered to Auto-Ordnance in the form of certified checks.[15]

At his Bronx warehouse, de Lacy and his associates unpacked the submachine guns, obliterated their serial numbers, and repacked them in large burlap shipping bags. On Sunday, June 12, the bags were loaded aboard a motor launch and carried down the Hudson River to Hoboken, where the collier *East Side* was tied up at Pier 2. The *East Side* was bound for Ireland by way of Norfolk, where it would take on a load of coal for the city of Dublin, but its engineers had joined other striking maritime engineers and walked off the job. This was the second time the ship had lost a crew to the strike, and the ship's owners regarded it as merely their good luck that almost immediately a young Irish chief engineer had showed up wanting to sign on. They regarded it as extraordinary luck when he also promised to furnish his own crew. The new chief engineer and his crew, likewise Irish, went aboard the ship on Saturday, June 11, and spent most of the next day industriously loading heavy burlap bags of "engine-room supplies" that were delivered alongside by a motor launch. In order not to arouse suspicion, the Irish crewmen left the bags littered about the deck until nightfall, then began slipping them below, one by one.[16]

If the bags did not arouse suspicion, the Irish crewmen did. The ship's chief steward had noted their unusual clannishness and had marveled at their unusual display of energy for a Sunday afternoon. Later, he noticed that the bags seemed to be disappearing. This aroused his suspicions even more, and when he stumbled onto one of the bags near his own quarters, with no one else around, he decided to investigate. Cutting a slit in the bag with his pocketknife, he peeked inside and into the muzzle of a gun.

While the steward was excitedly reporting his find to the ship's captain, the bag vanished, and a search failed to turn up any evidence of guns. The captain dismissed the matter as a case of "strike jitters" inspired by rumors of plots to blow up ships. Probably the steward had seen a bag of pipes.

At this point the plot still might have succeeded, had the nervous Irishmen not tried to get their guns off the ship. On Monday evening, just after dark, the mysterious motor launch reappeared, and the men on board told the ship's watch, in conspicuously Irish accents, that they had come to remove some supplies delivered to the *East Side* by mistake. But the watch, having heard the steward's gun story, wanted no strangers coming aboard his ship for any reason and backed up his feelings with a squad of crew members armed with clubs. The launch disappeared, and so did the Irish engine-room crew.

The incident with the launch and the disappearance of the Irish crewmen convinced the ship's captain that there was indeed some kind of devilment afoot. A more thorough search of the ship began the next day, and on Wedesday morning, June 15, the guns were found in their burlap bags, hidden under coal in the ship's bunkers.

The find consisted of thirteen bags containing 495 Thompson submachine guns, 1,854 magazines in all three sizes, plus extra parts, cleaning equipment, instruction booklets, and a case of .45-caliber ammunition. Three of the guns were completely assembled and ready for firing, which gave rise to newspaper speculation that the Irish crew members had intended to seize the ship and land the guns at some isolated place on the coast of Ireland.

De Lacy had put all his Tommyguns in one basket for several reasons. The *East Side*, Dublin-bound and strikebound, was the perfect smuggling vehicle. One large shipment would stand a better chance of getting through undetected than several small shipments. And once this new and deadly weapon *was* detected, either in shipment or in a guerrilla raid in Ireland, the United States government would doubtlessly throw an antismuggling fit to mollify the British. But the gamble failed, and with it Irish hopes of launching any spectacular offensive.

In a last-ditch effort to save the guns, "Frank Williams," accompanied by two Hoboken policemen, made his way to Pier 2 just as the seized guns were being loaded onto trucks. He had with him a warrant claiming the guns had been stolen from his warehouse a few days earlier by a rascal named Brophy; now that they had been found, he wanted them back. At the Hoboken police station, where the guns were taken, he pursued the matter further, but to no avail. United States Customs agents wanted clearer answers concerning his "import-export" business and the "tip" that led to his most timely arrival at Pier 2. "Williams" also chatted with reporters, some of whom remarked on his conspicuous brogue. Asked if he might just happen to be Irish, he replied, apparently with tongue in cheek, "Sure, you must be mishtayken!"[17]

From Hoboken, the guns went to a government warehouse and de Lacy went to Ireland.[18]

The smuggling attempt made front-page news across the country and remained on the inside pages of New York papers for many months as the government continued its investigation. The man handling the case for the Department of Justice was Assistant United States Attorney Roy C. McHenry, an eager but rather inexperienced prosecutor from the Attorney General's office. Assisting McHenry was a promising young man from the Bureau of Investigation, J. Edgar Hoover. After acquainting himself with the case, McHenry set out with great determination to nail Marcellus Thompson, Frank Merkling, the Auto-Ordnance Corporation, and its Sinn Fein customers with conspiracy charges, but succeeded only in getting himself fired.

From the Auto-Ordnance documents seized by federal agents, McHenry suspected that both Marcellus Thompson and Walter Morgan were aware of the smuggling plot, and that Frank Merkling was even more deeply involved. Company correspondence seemed to indicate that both Merkling and Morgan had a close working relationship with Ochsenreiter, and that Ochsenreiter was linked with one of the company's other Irish customers, John J. Murphy. On March 3, 1921, Merkling had wired Morgan in London:

> O APPREHENDS HIS PRINCIPAL MAY CANCEL ORDER FOR HUNDRED AS HE COULD FORFEIT DEPOSIT AND GAIN BY DEALING THROUGH M AT TWENTY OFF AFTER GOING INTO EVERY PHASE OF MATTER SUGGEST ADDITIONAL DISCOUNT OF FIVE PER CENT WITH OPTION ON DELIVERIES AS HERETOFORE PROMISED PROVIDED O SECURES ORDER FOR EIGHT HUNDRED ADDITION BY NEXT WEDNESDAY WIRE ME CARE O[19]

The "O" stood for Ochsenreiter, through whom Rorke had initially ordered a hundred guns. The "M" evidently stood for Murphy, who was the only other individual customer interested in buying Thompsons in any quantity at this time.

On March 4, 1921, Morgan wired his answer to Merkling, in care of Frank Ochsenreiter in Washington:

> GRANT EXTENSION OF OPTION ON EIGHT HUNDRED UNTIL WEDNESDAY NEXT AT ADDITIONAL FIVE PERCENT DISCOUNT WITH UNDERSTANDING THAT IF EXPECTED DIFFICULTIES DO NOT MATERIALIZE AND DEAL CLOSED BY O ON ORIGINAL BASIS THAT HE DOES NOT DEMAND THIS CONCESSION STOP HAVE WISE REPLY IMMEDIATELY OUR TELEGRAM YESTERDAY ASKING IF HE IS A MEMBER OF RESERVE CORPS AND IF HE WAS DECORATED[20]

Morgan seemed well acquainted with the finer points of the sale, as well as its "difficulties." His somewhat cryptic query about George T. Wise, Ochsenreiter's cousin and Auto-Ordnance's Washington salesman, may have been a routine company question, for Wise was in fact a former Army Air Service officer. Or it may have been a query as to Wise's affiliation with the Irish Republicans.

McHenry had no doubts that Merkling, at least, was in cahoots with Ochsenreiter and the Sinn Fein. He also had G. Owen Fisher's statement that Marcellus Thompson had told him to try to sell the submachine gun to the "Irish crowd." All told, he had a fascinating but largely circumstantial case against sixteen individuals, and he took it before a New Jersey federal grand jury in September 1921.

Before the grand jury had a chance to act on the case, Assistant Attorney General John W. H. Crim arrived in Trenton, fired McHenry, and let the jury go home. McHenry resented such treatment. Later he gave a reporter from the New York *World*[21] a story implying that the case was being hushed up because of the prominent persons involved, namely Marcellus Thompson, Thomas Fortune Ryan, and Ambassador Harvey. In response to the story, Crim defended his actions in a memo to the Attorney General. He described McHenry as "the least competent man I had ever seen sent out by this office to handle a case," avowing that McHenry had little knowledge of evidence and could never have proved his charges. He wrote that the United States Attorney in Trenton

> . . . called me on the telephone a day or two ago that he was not satisfied with [McHenry's] investigation, that he wanted to continue the matter over from this grand jury to another, and asked me if I could send him a man from Mr. Burns' department [the Bureau of Investigation] to do some very delicate work in this case. . . .
>
> The United States Attorney is handling this case in a very intelligent manner, and it may be that he will obtain sufficient evidence to justify the indictment of Col. Harvey's son-in-law and others. There has been no interference with this case since it was turned over to the United States Attorney. [He] has a free hand, he is under no restraint whatever by this Department, and he ought to be permitted to work out his situation in his own way.
>
> In conclusion permit me to say that I never heard the name of a single person involved in this matter until Mr. McHenry told me over the telephone. I did not know Col. Harvey had a son-in-law, I do not know Col. Harvey, and I have never talked to anyone with respect to this case outside the members of the Department of Justice.[22]

In January 1922 the Department of Justice returned the case to a Trenton federal grand jury. Three months later secret indictments were returned against Marcellus Thompson, Frank Merkling, the Auto-Ordnance Corporation, George Gordon Rorke, Frank Ochsenreiter, Larry de Lacy (under the name of Williams), and three of de Lacy's associates. All were charged with violating Section 13 of the Federal Criminal Code of 1922 prohibiting any "conspiracy to set on foot and provide the means for a military enterprise to be carried on against the territory of a Foreign Prince with whom the United States is at peace."

Thompson was arrested on June 19, 1922. The next morning the *New York Times* placed the story on its front page, with the embarrassing headline

HARVEY'S SON-IN-LAW
HELD IN ARMS PLOT

Released on $2,500 bail, Thompson refused to discuss the matter with reporters beyond insisting that he had no idea of how the guns could have gotten on board the *East Side*, and that "Of course we would not think of selling guns to persons we might even suspect of reselling them into the hands of enemies of constitutional governments."[23] George Harvey, at his post in London, simply refused to comment, and according to *Time* magazine, "went to his grave soft-peddling his interest in the company which had sold 750 [*sic*] choppers to the Irish Republican Army."[24]

Despite the sensation caused by the plot, the investigation, and the indictments, nothing ever came of the case. The British, who at first seemed eager to send prosecution witnesses and evidence that Tommy-guns had reached the I.R.A., later backed down. An *aide-mémoire* to Secretary of State Charles Evans Hughes in March 1922 said:

> . . . in view of the changed situation in Ireland, His Majesty's Government would greatly prefer not to proceed to produce the evidence specified. His Majesty's Government are, accordingly, anxious to ascertain unofficially whether the United States Government would be sat-

isfied to confine the case to the domestic issue involved in the illegal shipment of arms on a ship loading in a United States port.[25]

Nor did the Justice Department any longer have its star witness, salesman Owen Fisher. Already beset by personal problems, including the serious illness of his wife, Fisher had suffered a severe "nervous breakdown" upon learning that Thompson guns had been found aboard the *East Side*. He was on a sales trip at the time, and returned to New York to find that he had just been fired from his job with Auto-Ordnance, ostensibly for reasons of economy. After being questioned by federal agents, whom he impressed as sincere and cooperative, he suffered a complete collapse and was taken to the Kings County Hospital as insane. Ten days later he died of pneumonia.

After the return of the indictments, the government's attorneys reviewed the case and earlier court decisions interpreting the neutrality statute involved, and decided they lacked sufficient evidence to get convictions. The case was nol-prossed in 1923.[26]

Nor could the government keep the captured Tommyguns. The Justice Department discovered that the guns had been seized under wartime legislation that already had been repealed by 1921. On September 14, 1925, the department entered an order of discontinuance in the case of *United States v. 495 Thompson Machine Guns*, directing the New York Collector of Customs to deliver the guns to "Frank Williams, claimant and owner, or Joseph McGarrity, his duly authorized agent and attorney-in-fact."[27] McGarrity, a wealthy Philadelphia distiller, happened to be the leader of the radical faction of the Clan na Gael which opposed the Free State Treaty of 1921 and supplied arms to de Valera's irregulars during the Irish Civil War of 1922 and 1923. McGarrity accepted the submachine guns on behalf of de Lacy, then living in Dublin, and probably continued smuggling them to the "illegal" I.R.A. in the years to follow.

The Tommygun played a role in the Sinn Fein rebellion and in the Civil War that followed, but not the spectacular one now accorded it in Irish lore and legend. In fact, there is only one recorded instance of the gun's use prior to the July 9, 1921, truce—in an attack on a British troop train at Drumcondra on June 16.[28] The Thompson was employed more extensively in early 1922, in border raids against Ulster. In an attack at Clones Station in February Tommygunners killed four Special Constables and wounded eleven others.[29] But during the Civil War most of Ireland's Tommyguns were in the hands of the Free State Army, not the I.R.A.[30] A letter captured by Free State forces in the summer of 1922 revealed the intention of de Valera to obtain a new supply of Thompsons through Joseph McGarrity in the United States, but apparently few, if any, got through.[31]

Why the I.R.A. did not use what Thompsons it had prior to the 1921 truce is something that still puzzles old veterans of the rebellion. A few of the guns may have reached the rebels as early as April, and the fifty purchased by John J. Murphy seem to have reached Ireland during June and July.[32] It may have been that I.R.A. leaders did not want to alert the British to their new weapon, knowing there would be repercussions in the United States that might make it more difficult for de Lacy to smuggle in the big shipment. The Tommygun attack at Drumcondra did not take place until June 16, the day after de Lacy's guns were captured in Hoboken.

Although the Tommygun did not reach Ireland in large numbers or play a major role in the winning of Ireland's freedom, it was a novel enough and glamorous enough item of ordnance to earn itself a place in the nostalgic memoires of I.R.A. veterans. One such account describes the young rebel soldiers as adventurous, idealistic farm boys and students who "enjoyed the irresponsible romantic existence, the swagger which was made possible by a revolver or a rifle (thrice-happy the man who could carry a Thompson gun!)."[33]

The chorus of one old I.R.A. ballad goes:

> We're off to Dublin in the green, in the green,
> Where the helmets glisten in the sun,
> Where the bayonets flash and the rifles crash,
> To the echo of a Thompson gun.[34]

4

A Machine Gun for the Home

The ideal weapon for the protection of large estates, ranches, plantations, etc.
—*Auto-Ordnance advertisement, 1921*

The guns sold to the Irish rebels represented more than an embarrassing incident in early Auto-Ordnance history. They represented the one and only large order the company received throughout its early years.

From the start, Auto-Ordnance directed its strongest sales efforts at domestic and foreign military markets, hoping the submachine gun might "catch on" in military circles as the ideal all-purpose weapon for peacetime armies. Any small arm officially adopted by an army usually remained "standard" for many years, and sooner or later would be needed in great numbers.

In May 1921 General Thompson, whose role in Auto-Ordnance was mainly that of ceremonial head, sailed to London to demonstrate the gun for the British government. The company had been courting the British since December 1920, and prospects had seemed good enough to lead Auto-Ordnance to open negotiations with Birmingham Small Arms, Limited, with a view to manufacturing the Thompson gun in England under license. But while Thompson waited in London for George Goll to arrive with some sample submachine guns, news came in of the *East Side* smuggling plot, and British interest turned to suspicion. When the British further learned that the new American ambassador was related to Marcellus Thompson, suspected gunrunner, sales prospects vanished. George Goll recalls: "When we arrived at the British War Office, in a nice way they asked if we had already demonstrated the gun to the Irish. We were plenty embarrassed."[1]

From England Thompson and Goll went to Belgium, France, and Spain, where the gun was received with enthusiasm but not much else. Of the three countries, only Belgium seemed a likely market. In a demonstration at Camp de Beverloo, near Bourg-Leopold, George Goll sprayed a thousand rounds back and forth along a simulated enemy trench—a strip of brown paper fifty yards long—with excellent effect. The performance greatly impressed the high-ranking army officers present. General Bernheim, Inspector General of the Infantry, remarked that if he had had Thompsons in 1914, the Germans could never have taken Belgium.[2] The army indicated it was seriously interested in adopting the gun, and King Leopold authorized the purchase of a hundred for use in the Belgian Congo, but there the matter remained.

From other parts of Europe, marketing reports also were encouraging—at first. For a time Rumania seemed on the verge of adopting the gun and ordering a large number, and Czechoslovakia expressed serious interest in buying Thompson patent rights and putting the submachine gun into production at Skoda, the country's national armament firm. Poland, too, seemed more than routinely interested. But nowhere did negotiations go much beyond the talking stage.[3]

In Cuba, Mexico, and Central and South America the situation was about the same. Government and military officials turned out to watch elaborate demonstrations, applauded the gun's splendid performance, and then consigned it to the limbo of weapons "under consideration." A few small lots were ordered for trial purposes and for police work, but no sizable military sales developed.

During this period only one large arms sale came even near to materializing—a rather mysterious "large Far Eastern deal," apparently with China, for 50,000 guns plus accessories and ammunition. Company correspondence between April and June 1921 refers cryptically to a big "Shank deal" and also to an "Oxie matter," which seems to have been causing complications. On April 26 Marcellus Thompson wrote to sales manager Walter Morgan:

> We know you have been wondering what happened in connection with our trip to Washington on the S [Shank] deal. It sounds almost too good to materialize, but we came back with a tentative agreement on their part to purchase 50,000 Thompsons, 200,000 extra 20-capacity box magazines, 50,000 50-capacity drum magazines and 50,000,000 rounds of ammunition for a total of $14,600,000; 70% cash and 30% bonds. The tentative agreement includes an arrangement with Oxie and his associates whereby the net receipts by us under no conditions reach a sum of less than $140 cash per gun and the rest in bonds. All that now seems necessary is a further cable of approval from the Ruler, and an initial deposit of a few hundred thousand dollars in cash to our credit. The S group seems to think the matter is about consummated. We feel so encouraged that we are having Larkin dress up the agreement in legal form. Of course, it may all fall through but the scene shifts to New York at the end of the week when S and his associates come up to look over the agreement in the final shape and we hope to affix their signatures with the authority of the Ruler. We feel we have a fair chance of having "export restrictions" removed.[4]

On June 9, 1921, Frank Merkling wrote to Morgan:

> The Oxie matter has taken many queer turns since your departure. In fact, you would not recognize it as the same deal. The Chicago man developed into an awful bloomer—a regular Doctor Cook—but he had a valuable paper which became troublesome. It was not an asset after we found that he was not the party who could consummate the deal. On our trip to Washington which was made just about the time you were leaving to go abroad, we got in touch with the real power in this matter and have never let him loose. He is definitely committed to ten thousand and proportionate equipment and ammunition and assured me last week that he was arranging to finance the matter in the U.S.

It will come to a head the latter part of next week, and we feel extremely hopeful that there will be no further hitch. In any event, he is the real power in touch with the Ruler, with authority to sign contracts and issue notes in the Ruler's name, and to bind him in other ways. He has the broadest possible powers which have been ratified by the parliament of that country. . . . You can imagine what a delicate situation developed when we had to transfer in mid-ocean in a turbulent sea from the back of the Chicago man to this new man who turned out to be the real authority, but the change was effected without anybody falling overboard with the exception of the Chicago man.[5]

This fast-ripening arms deal came to the attention of the Justice Department when federal agents were investigating the Irish smuggling plot. Under questioning by J. Edgar Hoover, Merkling admitted that the company had friends in Washington working to obtain special legislation, in the form of an ambassadors' agreement, that would evade the existing embargo on arms shipments to China. This disclosure may have scuttled the deal; or it may have bogged down in a prices and terms dispute alluded to in other correspondence. In any case, nothing further came of the $14,600,000 deal that almost certainly would have guaranteed the financial success of the Auto-Ordnance Corporation.

If foreign military sales fell short of expectations, the once-promising domestic military market, when approached, vanished like a mirage. The Army paid the gun its highest compliments, then decided it was too big for a pistol, too small for a rifle, and therefore tactically valueless. The Navy and the Marines took the Army's word on this as a matter of small-arms policy. Only the National Guard in states troubled by "labor riots" purchased guns for other than testing purposes.

Compared to the Lewis light machine gun, which sold for about $650, the Thompson was a bargain at $225. About 1923 Auto-Ordnance reduced this price to $175, but the government still was not interested.

The disappointing response from the military was due in large part to the times. Depressed economic conditions in Europe, and in the

United States a wave of antimilitarism, had cut military budgets to the bone. An enormous amount of blood and money had been spent in making the world safe for democracy. Few countries had the funds, the inclination, or any pressing need to buy Thompson submachine guns.

But the main obstacle to the submachine gun's acceptance, and to the success of Auto-Ordnance, was a traditional one—military conservatism. The United States Army (and other armies as well) did not know quite what to make of a small arm that was not a pistol and not a rifle, and military men tended to reject it on grounds that it could do the job of neither. But the military everywhere understood and accepted the BAR, and this inspired Auto-Ordnance to develop the Thompson Gun, Model of 1923—a slightly modified version of the original submachine gun, disguised to look like an automatic rifle.

The Model 1923 Thompson was introduced that year with the usual fanfare and demonstrations. It had an unfinned barrel of fourteen inches instead of a finned barrel measuring ten; it had a durable horizontal forearm instead of the submachine gun's pistol-type foregrip, which might have broken off in combat; it had a bipod, a bayonet attachment, and a higher buttstock to facilitate firing from the prone position; and it was chambered for a new, specially developed Remington-Thompson .45-caliber automatic cartridge of greater range and power than the standard pistol round. The Auto-Ordnance catalog for 1923 listed it as the Thompson "Military Model," and the regular Model 1921 as the Thompson "Short Barrel Model," dropping the alien term "submachine gun" altogether.

But the gun was still a Thompson, made up from existing parts (except stocks and barrel), and the military was not fooled. After a series of futile demonstrations in the United States, Europe, and South America, the Model 1923 Thompson was abandoned as a total flop, and the term "submachine gun" restored to the Auto-Ordnance vocabulary.[6]

Nor did the submachine gun fare much better as a law-enforcement weapon, despite the most opportune timing imaginable.

The entire country was in a flap and the police in a quandary over the new-style "motorized bandits," who could hit a bank or a store or a payroll messenger and beat it quickly out of town, or even out of state. Occasionally the police would arrive on the scene in time to give chase, and the spectacular running gun battle between cops and robbers came into vogue. But without two-way radio, the pursuit was merely an auto-age version of the old outlaw chase on horseback, and the cop, hanging out the door of a careening Model T, had little chance of halting a higher-powered bandit car with only his service revolver.

Also vexing the police at this time was the increasingly serious, or at least increasingly publicized, problem of riot control. Postwar labor troubles and national anxiety over the "Bolshevist threat to America" made any noisy crowd seem on the verge of rioting and terrorism.

The ideal police weapon for dealing with either speeding bandit cars or rampaging mobs was, of course, a hand-held machine gun, according to Auto-Ordnance. For stopping armed robbers, the Thompson was extensively promoted as the "antibandit" gun, against which no getaway car would stand the slightest chance. For mobs, Auto-Ordnance mercifully offered a special short-range .45-caliber Peters-Thompson shot cartridge containing 120 pellets of No. 8 bird shot; it advertised the cartridges as "useful to authorities in dealing out a lesser degree of punishment. They allow serious occasions or disorders to be handled by officers of the law in the most humane manner possible."[7]

During 1920 and 1921 Auto-Ordnance had circularized city, county, and state police agencies and prisons, armored car companies, large banks, and private protective agencies. Even before the submachine gun went into production, prototypes were furnished to the police departments of Cleveland and New York to generate newspaper and magazine publicity. In *Scientific American*, a long article pictured two New York City policemen posturing menacingly with their new "pocket machine guns," and glowingly described the "wicked little submachine gun" as "interesting to those persons who wish to see the customary New York brand of gunplay made a little less one-sided." The writer, Army Captain E. C. Crossman, disparaged the average policeman's marksmanship as "something to make honest men turn pale, and the women and children duck for the subway." But he went on to predict that "the early future will see a happy coincidence of a policeman skilled in the pointing of the new weapon, and an automobile full of yeggs willing to engage in the customary running gun fight. The result will be the worst-shot-up assortment of crooks that has come to the attention of the coroner."

In equally tortured syntax, Captain Crossman assured his readers that it would be "no trouble whatever for one man firing the gun to sweep a street clear from curb to curb, but after all, its greatest strength lies in its moral effect. Killing many of the common American sort of mob is unfortunate unless the right ones can be selected for the slaughter." The author advised sparing the fools and concentrating fire on those vicious enough to try to stand up to such a weapon.[8]

In May 1922 Auto-Ordnance staged an impressive exhibition of the gun's antibandit qualities for a crowd of some two hundred newsmen and police and military officials at Tenafly, New Jersey. At a luncheon preceding the demonstration, John Thompson declared:

> Communities and their departments of police can no more counteract the organized banditry of today by the defense and pursuit methods of twenty years ago than the great war could have been won by the methods employed in the Spanish-American war. This weapon changes the odds which are now against the officers of the law and gives them instead a decided advantage over the bandit. I think you will agree, after seeing this gun in action this afternoon, that any man will think twice before going up against one.[9]

Then Thompson led his guests to the shooting range to witness a series of demonstrations. In one spectacular stunt, a policeman firing a submachine gun mounted on a motorcycle sidecar riddled the tires of an empty "getaway" car which had been cut loose from its tow. As the car thumped to a stop, two more Tommygunners moved in, firing both regular and incendiary ammunition. Judging from newspaper accounts, this last bit of showmanship had great spectator appeal—at least for reporters, who were enormously impressed that the antibandit gun would now permit police not only to shoot fleeing criminals but also burn them to a crisp. According to the *New York Herald*, "The three guns, pounding out steel jacketed .45-caliber bullets at the rate of 1,000 a minute, sliced the tires off the machine, cut through the spokes and chewed their way through the body, radiator, lamps, steering gear, tanks and woodwork." Then the incendiary bullets "converted a solid looking seven passenger touring car into a flaming mass of junk in less than a minute. Nothing in the machine larger than a mouse could have survived the fusillade."[10] And even the mouse, presumably, would have got it when the tank went up.

Three days after the Tenafly show, the motorcycle-mounted Tommygun was prominently displayed in New York's annual police parade up Fifth Avenue. News stories reported that the police were considering adopting the gun in a new "command of the road" plan. Fast sidecar motorcycles, equipped with Thompsons and two-way radios, would "scour the territory, affording little opportunity for the escape of automobile bandits."[11]

But law enforcement agencies did not buy enough antibandit guns to make much of a dent in the Auto-Ordnance inventory. By 1923, small numbers of Thompsons had been sold to the police departments of New York, Boston, and San Francisco, and to the state police of Pennsylvania, Massachusetts, West Virginia, Connecticut, and Michigan.[12] The Texas Rangers and the sheriffs' departments in mining and manufacturing communities also bought a few. But in most cases the gun sold less on its merits as an antibandit gun than on its power to intimidate striking workmen. Despite their initial enthusiasm, police officials in large cities seem to have had sober second thoughts about whether they should "sweep a street clear from curb to curb," bandits or no.

By 1925 about three thousand Tommyguns had been sold, but at a steadly declining rate, and it was painfully clear that the company's original marketing estimates had been far too optimistic in every area. Indeed, it was clear that sales expectations had been based largely on wishful thinking, inspired perhaps by the company's own extravagant claims.

As initially conceived, the submachine gun was purely a military weapon, and a highly specialized one at that. But by the time it went to market the trench fighting was long since over. To peddle the new gun in peacetime, it was necessary to think up something else it might be good for. With more enthusiasm than perspective, Auto-Ordnance decided that the submachine gun was, in fact, good for anything. Especially anything hard to hit, or that needed to be killed in quantity, or thoroughly intimidated. At practically any range. Auto-Ordnance claims could have been boiled down to a single slogan: Anything a gun can do, the Thompson can do better.

In publicity releases and advertisements the Tommygun was depicted as the world's most versatile all-purpose firearm. It was (without buttstock) a large machine pistol "which can be carried under the coat for instant use."[13] It was (with buttstock) a small machine rifle with terrible striking power and accuracy at all ranges up to five hundred yards. It could slice a speeding car to pieces with murderous cupronickel jacketed slugs, or calm a raging mob with the gentle patter of bird shot.

"The gun, it is held, is without equal for riot use and for the police in chasing thieves and other lawbreakers who attempt to escape in automobiles, and even an inexperienced man, it is claimed, can fire with the effect of an expert marksman, and moving targets can be hit with the ease that a fireman sprays a hose on a flame,"[14] an article in the *Army and Navy Journal* reported. Equally impressed by Auto-Ordnance claims, *Army Ordnance* declared that "the military possibilities of the gun for airplanes, protection of artillery and heavy machine guns in action, and for cavalry use, apparently are limited only by the ingenuity of the officers of those services."[15]

But the Thompson was even more versatile than that. One Auto-Ordnance advertisement portrayed a classic Tom Mix-type cowboy, complete with ten-gallon hat, wool chaps, and submachine gun, blazing away from the hip at a bunch of Mexican bandits—equally classic in their sombreros and bandoleers. Beneath the spectacle of rearing horses and toppling Mexicans, the text described the Thompson as "The most effective portable fire arm in existence. . . . The ideal weapon for the protection of large estates, ranches, plantations, etc. A combination machine gun and semi-automatic shoulder rifle in the form of a pistol."[16]

An even more dubious application of the Thompson was presented in the Auto-Ordnance scheme for arming aircraft—an idea that led to the building of one of the most extraordinary, ungainly, and useless air weapons ever to get off the ground. The World War had suggested the tactical possibilities of using airplanes against ground troops, either independently or in support of an infantry offensive. Existing warplanes, however, had neither the armor for low-level flying nor the armament to do much damage strafing. To remedy both these deficiencies, Auto-Ordnance teamed up with New York aircraft designer John M. Larsen to put together an armor-plated airborne contraption bristling with no fewer than thirty Thompson submachine guns. The plane used was Larsen's newly developed JL-12, a two-man, all-metal, low-wing, open-cockpit attack plane powered by a single 400-horsepower Liberty engine. Its gun arrangement consisted of twenty-eight Thompsons with hundred-round drums mounted in its belly—twelve pointed slightly forward, six directly downward, and ten inclined slightly to the rear—plus two guns mounted on swivels for the pilot and copilot. This marvelous and diabolical machine, its promoters claimed, could sweep down over a trench or a troop column at a speed of 140 miles per hour and lay waste to an area 12 yards wide by 440 yards long.[17] With its customary enthusiasm for any new weapon, and in its usual prose style, the *Army and Navy Journal* predicted that the Larsen plane would revolutionize modern warfare.

> Picture the reserve squadrons of attack planes that an army commander will want. For counter-attack purposes, no better instrument could be imagined. Tired troops holding occupied trenches after a successful attack will have about as much hope of successfully doing so against the withering fire of a counter-attack from above as the smouldering ember that breaks ablaze in the prairie fire of doing any further damage in the midst of a charred acre of ground.[18]

The plane was submitted to the Army for testing in December 1921, and it flew. But its four hundred pounds of armor plate and five hundred pounds of guns and ammunition, the test pilot reported, gave it the flight characteristics of a flat rock. And when the copilot unleashed its enormous fire power by means of levers in the cockpit, the guns began jamming on their own deluge of ejected shells.[19]

Rivaling the cowboy and the airplane ideas was one other suggested use for the submachine gun. The *Army and Navy Journal*, in another article doubtlessly based on a press release, advised that the Thompson "can be kept in the home as a protection against burglars."[20]

A company that could fancy a cowboy mowing down bandits, or envision a householder pouring machine-gun fire into his darkened dining room in defense of the family silver, might well have misjudged its markets. But the submachine gun was legally available to anyone, and lack of police and military interest made it, by default, a civilian weapon. And so it came to pass that the Thompson—manufactured in peacetime, sold on the commercial market—was, in a sense, a machine gun for the home.

5

The Gun That Made the Twenties Roar

On the Side of Law and Order.
—*Motto of the Auto-Ordnance Corporation*

At the Auto-Ordnance demonstration near Tenafly in 1922, observers predicted to newsmen that the new submachine gun was destined "to reform or remove bandits instantaneously."[1] It evidently occurred to no one at the time that a bandit might himself acquire a submachine gun and, if not reform, at least remove his pursuers. Auto-Ordnance was dimly aware that a Tommygun in the wrong hands would not be a good thing. Its earliest catalogs carried the brief note: "Thompson Guns are for use by those on the side of law and order and the Auto-Ordnance Corporation agents and dealers are authorized to make sales to responsible parties only." The only other expression of concern came from British arms expert H. B. C. Pollard writing in the *Saturday Evening Post* in 1923. Alarmed at indications of world rearmament and international gunrunning, he considered submachine guns in particular a potential source of trouble.

> Except as an arm for trench warfare or semimilitary police forces having to deal with armed risings, it is difficult to see what honest need they can meet; yet we are faced with the fact that they exist and have been manufactured in quantity and are on the open market for anyone who wants to buy them. Here, one would say, is an arm that is useless for sport, cumbrous for self-defense and could not serve any honest purpose, but which in the hands of political fanatics might provoke disaster.[2]

Even this writer did not recognize the Tommygun's potential as a weapon of criminals, although the country at this time was going through one of its periodic campaigns to control firearms.

With the increasing national crime rate following World War I, state and local authorities began enacting or stiffening laws regulating the sale and possession of firearms. New York City had led this movement in 1911 with its Sullivan Law, which prohibited even the ownership of a handgun without a license. Other cities did not go so far as to prohibit ownership, but many passed laws requiring licenses to purchase and permits to carry. These laws, however, were aimed chiefly at the celebrated "two-dollar pistol"—the hundreds of thousands of cheap nickel-plated revolvers bearing such quaint names as Tramp's Terror, Bang Up, and Parole which had been enlivening Saturday nights in America for a quarter century. It was the lawmakers' objective merely to get these "suicide specials" out of pockets and off the streets and to make them a little less available to criminals.

The new pistol laws often created a paradoxical situation. In some cities, including New York and Chicago, a person could not buy even a .22 target revolver without a license and without registering it, but he *could* buy, no questions asked, a Thompson submachine gun. Machine guns traditionally were large, heavy military weapons of no interest whatever to civilians, criminal or otherwise, and there never had been any need to outlaw them. Only "concealable weapons" seemed to need regulating, and the Tommygun, despite Auto-Ordnance advertising claims, did not come under that classification.

A few Thompson guns were in the possession of criminals as early as 1924, and probably earlier. Some East Coast rumrunners used them for protection against hijackers, who most likely had Thompsons also.[3] But since these parties normally avoided shoot-outs with the Coast Guard and confined their own battles to the privacy of the high seas, the Tommygun remained virtually unknown to the public until it was finally discovered by Chicago gangsters in the winter of 1925–26.

National prohibition profoundly changed the form and style of American crime. Large cities already had "gangs," the successors to nineteenth-century neighborhood and ethnic groups that had originally banded together for social reasons but later found profitable ways to employ themselves collectively as strong-arm squads in ward politics or in the "protection" of local vice operations. Because they already possessed manpower, organization, and political and vice connections, these gangs were well equipped to help relieve the distress caused millions of Americans by the Volstead Act after January 16, 1920. The enormous profits in bootlegging permitted the more ambitious gangs to grow into giant business operations. Bootlegging itself became a major national industry. Soon Americans learned some new words: "gangster," "racketeer," "organized crime."

For the first two or three years, organized bootlegging developed in a more or less peaceful and orderly manner, particularly in Chicago, where underworld leaders and ward politicians had long understood one another. Not only were there plenty of profits for all concerned, but the city's gang leaders at an early date had met and divided Cook County equitably among themselves in order to forestall competition. By 1923, however, Chicago's bootlegging gangs had so increased in both size and efficiency that supply began to exceed local demand, giving rise to territorial disputes. Lacking other means of resolving their differences, rival gangs began resorting to murder as an instrument of underworld policy. Shootings quickly escalated into full-scale "beer wars," and Americans learned some more new words: "Tommygun," "chopper," "typewriter," "Chicago piano."

The submachine gun was an ideal gangland murder weapon. It had all the virtues—light weight, concealability, tremendous fire power—attributed to it in Auto-Ordnance advertising. And it had other virtues as well. A gunman using a pistol had to move in dangerously close to his target, and if the first bullet failed to do its work, the target usually shot back; if the target was important enough to have bodyguards, a pistol was out of the question. A gunman with a 12-gauge sawed-off shotgun lacked nothing for deadliness, but he, too, was required to work at short range, and buckshot sometimes failed to penetrate heavy doors or automobile bodies shielding the intended victim. But a competent Tommygunner could get a kill (sometimes several) at any reasonable range, or from the safety and anonymity of a speeding car, and a machine gun had the bonus feature of deterring counterfire from the victim's bodyguards. They either went down with their employer or dived for the nearest cover.

The Thompson made its first appearance as a Chicago gangster weapon in the fall of 1925, following a period of deteriorating relations among the city's bootlegging factions.[4] In 1920 Johnny Torrio had succeeded the late (thanks to a Torrio gunman) Big Jim Colosimo as vice and beer king of the West Side, and it was at his urging that the city's gang leaders, in 1921, joined in a "union of each for the good of all."[5] Torrio, who was ably assisted by his chief lieutenant, Al Capone, remained supreme on the West Side. The rich North Side went to Dion O'Banion, Irish ward politician and flower shop proprietor. His second-in-command was Hymie Weiss, who reputedly invented the "one-way ride" and coined the phrase. The South Side was divided among several smaller gangs, the most prominent of which was the one led by Frank McErlane and Polack Joe Saltis.

All went well until 1923, when a former South-Sider, Spike O'Donnell, came home from prison to learn that his four fainthearted brothers had allowed themselves to be gerrymandered out of business when the redistricting took place. Led by Spike, the O'Donnells boldly moved into territory allotted to Saltis and McErlane; the shooting started; and the city-wide truce collapsed. North-Sider O'Banion not only doublecrossed Torrio in a brewery swindle, but began making disparaging remarks about Torrio's low-class Sicilian friends. In November 1924 Torrio dispatched the offensive florist at his North State Street flower shop in the famous "handshake" murder. Two months later Hymie Weiss avenged his former boss by having Torrio shot. Torrio survived his serious wounds, but retired to the safety of New York, leaving his West Side operation in the capable hands of Al Capone.

By the middle of 1925 the fighting had become general, with murders averaging about one a week, and woundings and attempts occurring almost daily. This was the beginning of the "Battle of Chicago," and gangland tacticians soon discovered a new and powerful offensive weapon in the Thompson submachine gun.

The Tommygun's gangland debut was inauspicious by later standards, and unsuccessful as well. The Saltis-McErlane gang introduced it in a determined effort to rid their territory of the O'Donnell brothers, who were becoming more than a nuisance. But at first the McErlane gunners showed little aptitude with their new weapon.

On the evening of September 25, 1925, Spike O'Donnell was standing at the corner of 63rd Street and Western Avenue, the busiest intersection in Chicago's West Englewood district, when he heard someone in a car shout, "Hey, Spike! Come here!"[6] O'Donnell dropped to the sidewalk as a terrific blast of gunfire took out the windows of the drugstore behind him. Unnerved by the experience, O'Donnell went into the drugstore and asked for a drink of water.

Chicago newspapers duly recorded another skirmish in the South Side beer war, but attributed the unusually large number of bullet holes to "shotguns" and "repeating rifles." The police likewise were impressed by the volley. O'Donnell, however, declined to discuss the matter beyond conceding that someone had indeed tried to shoot him.

His second time out, McErlane at least managed to score with his new gun, and get the gun mentioned in the papers. On the Saturday evening of October 4 a black limousine drove past the headquarters of the Ralph Sheldon gang, an O'Donnell ally, spraying it with bullets. Three of the shots struck and killed Charles Kelly, one of Sheldon's men, who was standing on the sidewalk. Two others wounded Thomas Hart, inside the building. This time the police figured out what all the bullet holes meant. According to the Chicago *Daily News*:

A machine gun, a new note of efficiency in gangland assassinations, was used to fire the volley from the black touring car, killing one man and wounding another in front of the Ragen Athletic club . . . at 5142 South Halsted St. last Saturday night.

Captain John Enright of the stockyards police said today his investigation satisfied him that a machine gun had been used, and that the same gun had been used in an attack on Spike O'Donnell at 63rd St. and South Western Ave. . . .

"The bullets were fired from a machine gun," Captain Enright said, "because we find that more than twenty bullets were fired into the club house. Witnesses say they came in too rapid succession to be revolver shots. However, we are basing our theory on something more that. . . .

"So far as their being bullets of the same kind as were fired at 'Spike' O'Donnell, I compared them with bullets which Englewood police picked up at the time and they are identical."[7]

Despite the machine gun, the story was buried on page 5. The wire services picked it up as a short human-interest item—the latest wrinkle in Chicago gangland warfare. But that was all. The papers paid even less notice when, not quite two weeks later, on October 16, the persevering McErlane riddled Spike O'Donnell's car on 95th Street near Western Avenue, wounding his brother Tommy.

If McErlane's early attacks were poorly reviewed by both the police and the press, he finally attracted attention as Chicago's Tommygun pioneer on February 9, 1926. The shooting, despite the usual poor marksmanship displayed, earned the *Chicago Tribune*'s front-page banner headline the following morning, and started the Thompson on the road to ill fame.

MACHINE GUN GANG SHOOTS 2

Thirty-seven bullets from a light automatic machine gun were poured into the saloon of Martin (Buff) Costello, 4127 South Halsted street, last night, by gangsters striving to assassinate two rivals for the highly profitable south side traffic in good beer.

Both men were wounded. William Wilson, 329 South Leavitt street, was shot in the head and probably fatally wounded. John (Mitters) Foley, 2838 Wallace street, vice president of the Ice Cream Wagon

Drivers' union, beer runner, and one time stickup man, was struck in the forehead, but was not seriously injured.[8]

The next day it was reported that Captain John Stege, Deputy Chief of Detectives, had expressed concern that "McErlane and Saltis have one of these guns" and had said that he would request similar weapons for the police. "It is imperative that if the police fight these men that they be armed accordingly."[9]

Although Frank McErlane introduced it, the man who made the Thompson famous as a gangster weapon was Al Capone, the Horatio Alger character of American crime. Starting as a lowly bouncer in one of Johnny Torrio's brothels, he quickly worked his way to the top, to become president of a multi-million-dollar booze and vice empire when Torrio retired in 1925. With his lofty position, however, came new problems—mostly in the form of stiff competition from the North-Siders. Hymie Weiss wanted a generous chunk of the West Side, and wanted Al Capone dead.

The Costello saloon machine-gunning made headlines on the morning of February 10. That same day, as if stirred by the thought that he was lagging behind in the underworld arms race, Capone went to the hardware and sporting goods store of Alex Korecek and placed a rush order for three Thompsons.

For openers, he killed three men and wounded two others in the first sensational "machine-gun massacre" of the Twenties. The occasion was an attempt to dispose of a certain James Doherty, a bootlegger allied with the North-Siders, who was poaching in Capone's home territory of Cicero. About dusk on Tuesday, April 27, 1926, Doherty and several other men parked their car in front of the saloon of Madigan and Wendell at 5613 West Roosevelt Road. As they were climbing out, the traditional black limousine roared by, Tommy-guns blazing from the window. An elderly woman who lived above the saloon, and who happened to be looking out her window at that moment, told police, "It was daylight still and I saw a closed car speeding away with what looked like a telephone receiver sticking out of the rear window and spitting fire."[10] Reporters counted more than a hundred bullet holes in the car and in the saloon.

The three men who died in the shooting were bootlegger James Doherty, a minor politician named Thomas Duffy, and William McSwiggin, the well-known Assistant State's Attorney for Illinois. It was first assumed that the gangsters had been gunning for the twenty-six-year-old "hanging prosecutor," as newspapers called him, and this idea shocked the entire country. Bootleggers were only supposed to shoot one another. But McSwiggin's death probably shocked Al Capone as well, for the police concluded later that Capone was after Doherty, and that it was simply McSwiggin's bad luck to be standing too close. Why the Assistant State's Attorney was fraternizing with bootleggers, no one could explain.

This was Capone's first Tommygun raid, and (according to the *Tribune*) police believed that he handled the gun himself "in order to set an example of fearlessness to his less eager companions."[11] If so, he was the cause of his own undoing, for the police, responding to public indignation, raided Capone's headquarters in Cicero. There they seized the ledgers that, five years later, in 1931, federal tax investigators would use to convict him of income tax evasion.

The machine-gun slaying of McSwiggin made headlines in New York the following morning and threw Auto-Ordnance into a turmoil. John Thompson was distressed. At the company's office he paced the floor expressing alarm and bewilderment. "What can we do?" he kept saying.[12]

The possibility, which was more a probability, that his submachine gun might one day become a weapon of criminals apparently had never occurred to him. Nor did it ever cease to trouble his conscience that the gun bearing his name became notorious as a murder weapon.

The McSwiggin murder likewise troubled Marcellus Thompson, partly for business reasons. After the Irish gunrunning scandal, Auto-Ordnance did not need any testimonials from Al Capone. The day the McSwiggin news reached New York, Marcellus Thompson boarded a train for Chicago to offer his assistance to the police, who may not have wanted it.

The murder weapon had been found; the killers had thrown it out of the car window while making their escape. But its serial numbers were thoroughly obliterated, and the police had let the matter go at that. Neither the police nor the gangsters were aware that every Thompson gun had a secret serial number stamped on the front of the receiver, concealed by the foregrip mount. Marcellus Thompson offered to trace the gun by means of its secret number and eventually did so, but without much help from the authorities. He told his story later to *Collier's*:

> I trailed this gun down by its number, and found that it had gone, through a prominent sporting-goods store in Chicago to a foreigner who owns a sporting-goods store in "The Valley," in Chicago.
>
> As soon as I got the address I got into a car and hurried out to the store. I felt that I was taking my life in my hands because I was sure that the gangsters knew I was in town and was trying to track down the gun. One automobile loaded with gangsters hovered around my car and passed us several times. I haven't the slightest doubt the men in this car were armed.
>
> The sporting-goods man admitted to me that he had sold one of my guns to someone—he didn't know who it was. Well, I hurried back to police headquarters and told the detectives what I had discovered. They went out and got the little dealer and brought him in. For three days they treated him so roughly that I began to pity him. They would browbeat him at conferences, and even the newspaper reporters would poke him in the ribs and say, "Come across with that story, ---- you."
>
> I went to the state's attorney's office too with the facts. William McSwiggin was an assistant state's attorney, and I wanted to be sure that the facts about the selling of the gun would get into the hands of State's Attorney Crowe. It was pretty hard to get to Mr. Crowe. I practically had to force my way in to him, and I told him what I had discovered about the dealer in arms.
>
> I had done all that I could do, and so I went back to New York.[13]

Police continued to question the dealer, Alex V. Korecek, who by this time was devoutly wishing he had never gone into the gun business. He admitted selling three Thompsons to a group of men who first came into his shop on February 10. From his reluctant

John Thompson, 1921. (George Goll)

Oscar V. Payne, about 1920. (Oscar Payne)

Theodore H. Eickhoff about 1935. (T. H. Eickhoff)

George E. Goll, in 1921. (George Goll)

Auto-Ordnance staff picnic near Painesville, Ohio, 1918. Theodore Eickhoff is at the far left. Standing nearest him are Elmer Koenig and George E. Goll (behind man with rifle). At far right is Oscar V. Payne, with Fred Deertz standing next to him. The others are not identified. (Oscar Payne)

Thomas Fortune Ryan.
(Wide World)

President Harding (left) and
Ambassador George Harvey. (UPI)

Marcellus H. Thompson and his
first wife, Dorothy Harvey, at their
wedding in 1914. (Evelyn Adams)

Marcellus H. Thompson, about
1930. (Thomas Kane)

Model 1 Thompson Autorifle prototype after it exploded during a test in August 1917. (T. H. Eickhoff)

First Blish test mechanism set up in the Auto-Ordnance experimental shop in Cleveland, 1917. (T. H. Eickhoff)

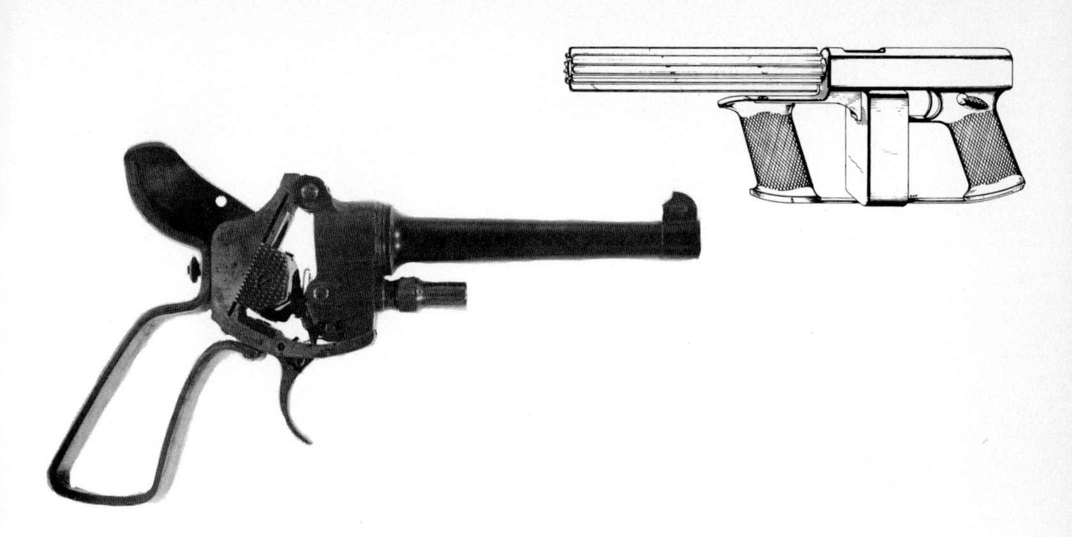

Above: Drawing of the submachine gun as it was originally envisioned by Oscar Payne in the fall of 1917. Left: Experimental single-shot handgun built by John Blish about 1915 to demonstrate the Blish locking system. (T. H. Eickhoff, George Numrich)

Auto-Ordnance machine shop and testing room in the Sabin Machine Company building in Cleveland, 1918. On the left is the testing platform and steel pipe leading to a bullet trap; on the right, behind a protective wooden screen, is a spark chronograph invented by Oscar Payne to measure the motion of the Blish lock in relation to the motion of the bullet. (T. H. Eickhoff)

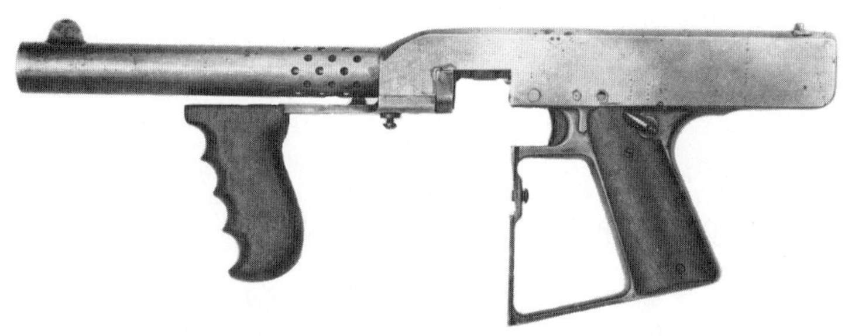

Above: "Persuader," equipped for belt feed. *Below*: "Annihilator," first working prototype, Serial #1. (Numrich, Nelson)

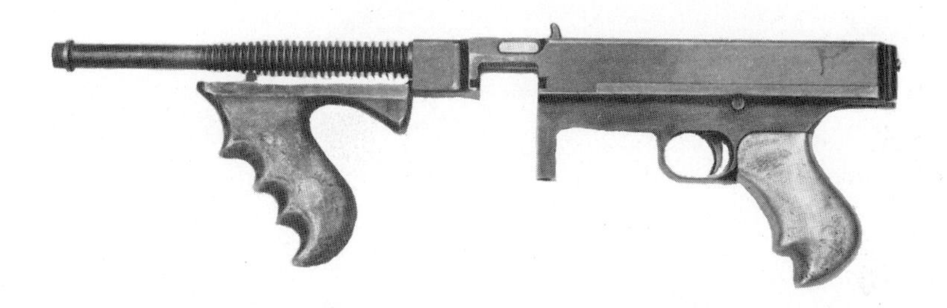

Prototype submachine guns. *Top:* Model 1919, Serial #2. *Above:* Model 1919, Serial #8, experimental 11-piece gun. *Right:* Later Model 1919 prototype with round actuator knob, but still lacking sights and buttstock. (Numrich, Nelson, Oscar Payne)

Type L 50-round drum magazine and Type XX 20-round box magazine. (George Goll)

Italian Revelli 9mm twin-barreled machine gun. (U.S. Army)

German Bergmann Machinen Pistole 18. (U.S. Army)

Submachine-type weapon built experimentally by Hiram Maxim. (Author)

Thompson Autorifle "P," Model of 1920. (Oscar Payne)

Experimental .22-caliber Thompson Autorifle. (Oscar Payne)

Demonstration of the prototype submachine gun in 1920, probably in Cleveland (Evelyn Adams)

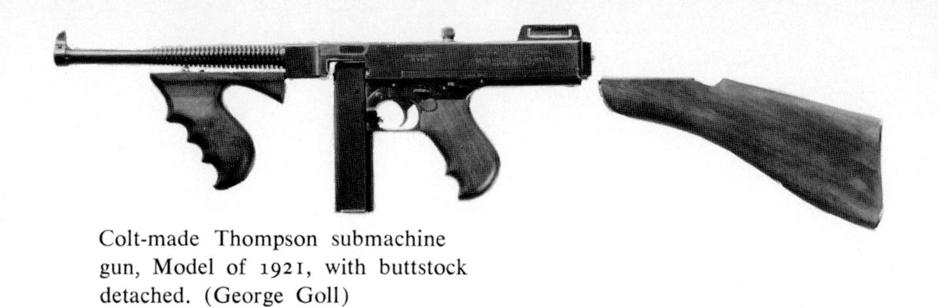

Colt-made Thompson submachine gun, Model of 1921, with buttstock detached. (George Goll)

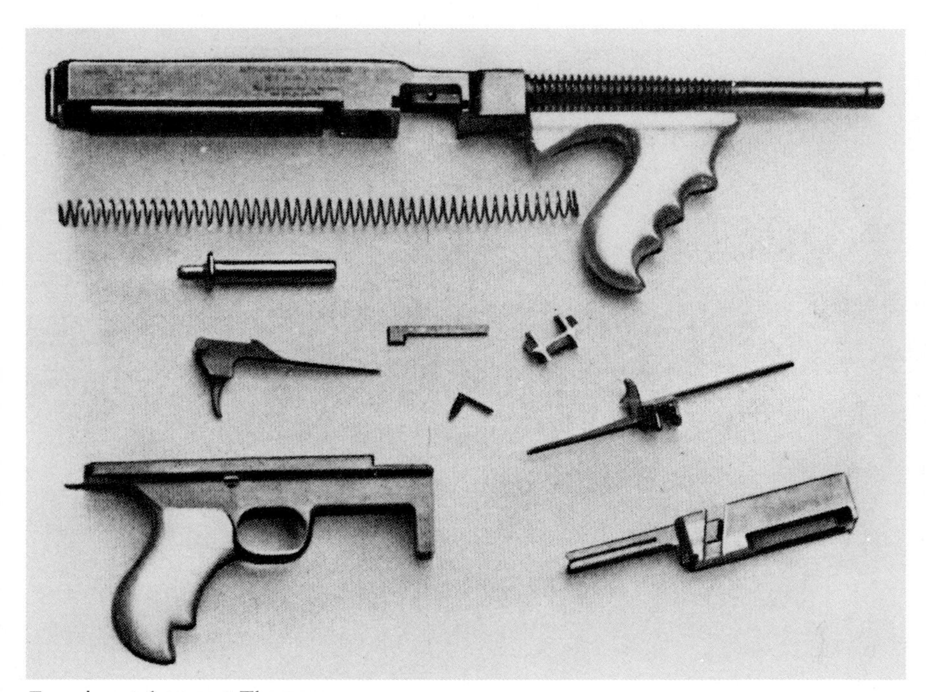

Experimental 11-part Thompson.

Model 1921 Thompson with Type C 100-round drum mounted on police motorcycle sidecar. (George Goll)

New York City Police "control of the road" plan in the early 1920s featured motorcycles equipped with a submachine gun and portable radio receiver. (Oscar Payne)

The Thompson Submachine Gun

The Most Effective Portable Fire Arm In Existence

THE ideal weapon for the protection of large estates, ranches, plantations, etc. A combination machine gun and semi-automatic shoulder rifle in the form of a pistol. A compact, tremendously powerful, yet simply operated machine gun weighing only *seven* pounds and having only *thirty* parts. Full automatic, fired from the hip, 1,500 shots per minute. Semi-automatic, fitted with a stock and fired from the shoulder, 50 shots per minute. Magazines hold 50 and 100 cartridges.

THE Thompson Submachine Gun incorporates the simplicity and infallibility of a hand loaded weapon with the effectiveness of a machine gun. It is simple, safe, sturdy, and sure in action. In addition to its increasingly wide use for protection purposes by banks, industrial plants, railroads, mines, ranches, plantations, etc., it has been adopted by leading Police and Constabulary Forces, throughout the world and is unsurpassed for military purposes.

Information and prices promptly supplied on request

AUTO-ORDNANCE CORPORATION

302 Broadway *Cable address: Autordco* **New York City**

Auto-Ordnance advertisement, 1920-21. (T. H. Eickhoff)

Marcellus H. Thompson presenting prototype submachine guns to Captain Charles Schofield, New York Police Department, in the fall of 1920. (UPI)

In 1925 the N.Y.P.D. inaugurated its Emergency Service with special trucks equipped with rescue and riot-control equipment, including Tommyguns. (UPI)

John and Marcellus Thompson, 1920. (UPI)

THE WORLD: TUESDA

ACCUSED GUN RUNNERS AND THEIR WEAPON

SOLDIER of IRISH REPUBLICAN ARMY with a THOMPSON SUB-MACHINE GUN

FRANK OCHSENREITER · Col. MARCELLUS H THOMPSON · GEORGE G. RORKE

EIGHT INDICTED

ALL ARMS TREATIES

New York *World* newspaper picture showing Frank Ochsenreiter, Marcellus Thompson, George Gordon Rorke, and an Irish Republican Army soldier with one of the submachine guns smuggled to the Irish rebels.

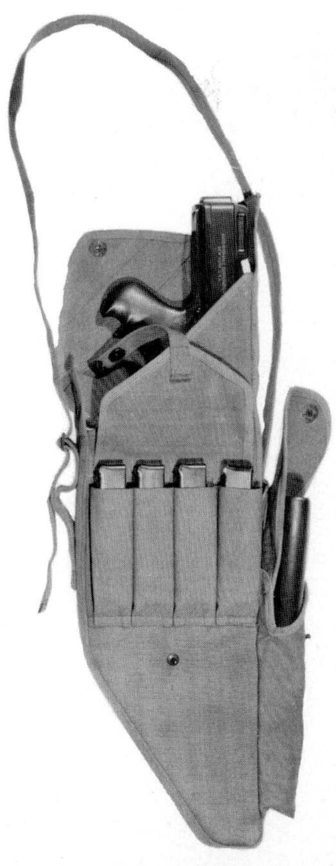

Canvas scabbard designed to hold submachine gun, buttstock, and magazines. (George Goll)

John Thompson posing with a group of Belgian military officials at Camp de Beverloo in 1921. (George Goll)

Thompson, far right, watching as George Goll demonstrates the submachine gun for Belgians, 1921. (George Goll)

Auto-Ordnance sales representative demonstrating a Model 1923 Thompson for Augusto B. Leguia (hand outstreched), President of Peru, about 1923. (George Goll)

George Goll firing the submachine gun at Camp Bisley in England, 1921. (George Goll)

Infantryman with Model 1919 prototype "trench boom" as pictured in an early Auto-Ordnance promotional booklet. (T. H. Eickhoff)

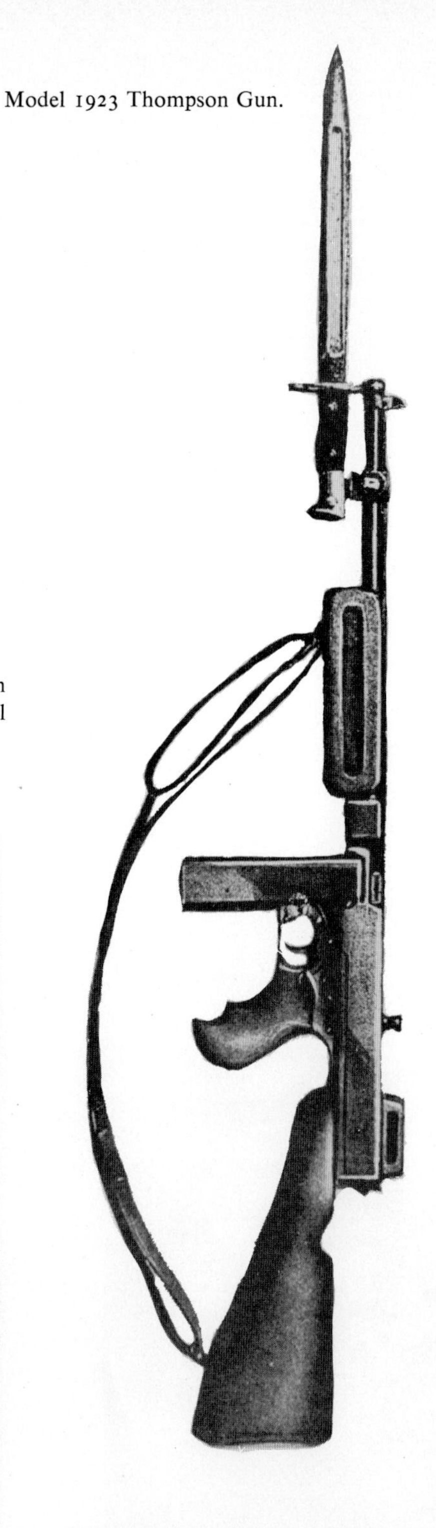

Model 1923 Thompson Gun.

Handbook of the
Thompson Submachine Gun
Model of 1921

Auto-Ordnance Corporation
302 Broadway
New York N.Y. U.S.A.

Original Auto-Ordnance handbook

descriptions, police decided that one of the men was Capone himself. Having admitted this much, Korecek decided that his chances of survival were nil and begged to remain in jail. A reporter described him as "speechless with fear." On May 1 a writ of habeas corpus mysteriously was filed for Korecek's release. Asked by reporters who his unknown friend was, Korecek replied, in so many words, "I don't know, but I know he isn't any friend of mine!"[14]

Korecek managed to survive both the cops and Capone, and Capone of course survived the police investigation.

With Capone in possession of machine guns, Hymie Weiss and his North-Siders lost no time in obtaining Thompsons of their own and putting them to work. Weiss had inherited North Side leadership following the murder of Dion O'Banion late in 1924, and since that time had spared no effort in trying to avenge his slain comrade. Dozens of attempts on Capone's life failed, and with each failure Weiss would turn to shooting at Capone affiliates, especially at the Genna brothers. Three of the brothers were dead and three in hiding when finally, in September 1926, Weiss resolved to hit Capone with everything he had. He failed, as usual, but did stage a tour de force never to be equaled in the annals of crime—a veritable extravaganza of shooting, known as the Siege of Cicero.

On the morning of September 20 Capone and some of his men were taking their ease in the coffee shop of the Hawthorne Hotel, the gang's headquarters, when their ears picked up the ominous rattle of a submachine gun coming from somewhere down the street. Capone rushed toward the window to see what the shooting was about, but a quick-witted bodyguard threw him to the floor. The shooting, predictably, was about Capone—intended to draw him and his men to the windows. For now moving slowly along Cicero's busy 22nd Street was a procession of about ten touring cars and limousines, bristling with machine guns, shotguns, pistols, and rifles. Weiss gunmen poured more than a thousand bullets into Capone's headquarters, but succeeded only in wounding a bodyguard, injuring a bystander with flying glass, and ruining the hotel.

Weiss's Cicero performance was spectacular, as befitting a valedictory. Twenty-one days later, on October 11, Weiss and four companions stepped out of their car in front of Weiss's office on North State Street and into a barrage of machine-gun and shotgun fire from a nearby second-story window. Weiss died with ten wounds and his bodyguard with fifteen; the other three were each hit several times but eventually recovered.

The killing of Weiss led to a temporary truce as the gangs stopped to tote up their losses and reflect on the wisdom of Torrio's early plea for peaceful coexistence. Promiscuous killing was bad for business, bad for employee morale, and bad public relations, for civic groups were beginning to agitate for an end to the gang rule that was making the country's second largest city a spectacle in the eyes of the world. The result was a formal peace conference at which Cook County was again equitably divided among the West-Siders, South-Siders, North-Siders, and the smaller gangs. Thus relative peace settled over war-torn Chicago for several months. Shootings did not resume on a large scale until nearly the end of 1927.

But if Chicago began to enjoy a period of unprecedented quiet, other parts of the country were now becoming acquainted with the Tommy-gun. In southern Illinois the gangs of Charles Birger and the Shelton brothers were engaged in a full-scale war with machine guns, bombs, armored trucks, and airplanes which eclipsed even Chicago's war for a time.

On October 26, 1926, the Sheltons ambushed and machine-gunned two Birger gang members on a country road near Herrin, Illinois. Not quite two weeks later, on November 7, four carloads of Birgers armed with rifles and submachine guns invaded the Shelton stronghold of Colp, killing the mayor and another man and wounding the chief of police. In retaliation for the shooting of one of their mayors, the Sheltons, five days later, hired an out-of-work barnstormer to drop three dynamite bombs on Charlie Birger's fortified estate near Marion. The bombs failed to explode, and on December 12, Birger gunmen went to West City and killed another Shelton mayor.

Skirmishing continued until January 9, 1927, when the Sheltons, out of patience and nearly out of mayors, mounted an all-out attack on the Birger estate. In the early hours of that Sunday morning residents in the area awoke to the sounds of machine-gun fire, exploding bombs, and (some claimed) zooming airplane engines. After the Sheltons pulled back, the police found nothing left of the Birger estate but a bomb- and fire-gutted ruin and several charred bodies, none of which was Charlie Birger's.

In city after city the "chopper" became standard gangland issue as the idea moved outward from Chicago. On February 25, 1927, Philadelphia experienced its first machine-gunning when local gang leader William Duffy fell critically wounded with six bullets fired from a speeding car as he stepped out of a nightclub. The attack killed the doorman and wounded Duffy's bodyguard, and was probably the work of rival bootlegger Max "Boo Boo" Hoff. Hoff, police discovered, for whatever it was worth, had recently laid in a supply of Thompsons and bulletproof vests, ordering them through a cooperative dealer in military supplies. In the same city two months later, on April 13, a Tommygunner in the traditional limousine killed two and wounded two others in a running battle with the occupants of a roadster.

In 1928 the Tommygun made its first violent appearance in New York City. Frankie Yale was an old friend of Capone's and for years a loyal, dependable purchasing agent who kept Capone supplied with the best booze smuggled into Long Island. In fact, he had been so dependable that when a number of Chicago-bound shipments were mysteriously hijacked Capone began to suspect treachery. He sent a spy to Brooklyn to investigate, and the spy confirmed Capone's suspicions by getting himself shot. On Sunday afternoon, July 1, 1928, as Yale was driving along 44th Street in the Homewood section of Brooklyn in his new Lincoln, a Nash carrying four men and two submachine guns pulled alongside. Yale died at the wheel, and his car, struck by a hundred bullets, plowed through some small trees and a hedge and into a house.

Meanwhile, back in Chicago, the truce of 1926 had long since collapsed, and Capone's supremacy was again being threatened by the ambitious North-Siders. During 1928 Capone lost one after another of his best men to enemy snipers, and he himself was the intended victim of several plots, including two second-story-window machine-gun ambuscades. By the end of the year, having lost some fifteen men during the previous three months, Capone decided that he had endured the unruly North-Siders long enough. Nor would it do merely to shoot their current leader, George "Bugs" Moran, when there always was a number-two man to take over. This time, Capone would wipe out the entire gang.

If Capone in the past had executed some fairly artful murders with the submachine gun, the project he now had in mind was a mural by comparison. Whether by coincidence or design (for Capone may have appreciated the irony), the slaughter took place on February 14, 1929—St. Valentine's Day. It was a masterpiece of planning, timing, and treachery, and, for murderousness, the Tommygun's finest hour.

To organize the project Capone appointed "Machine Gun" Jack McGurn, a talented young man who divided his time between professional golf and professional murder. McGurn's first step was to hire a free-lance hijacker, furnish him with a truckload of Old Log Cabin whisky supposedly stolen from Capone, and instruct him to sell it to Moran at an attractive price with the promise of another truckload later. All went according to plan. Moran bought the liquor and agreed to meet the second truck personally and pay for it with cash on February 14. The delivery would be made to the S. M. C. Cartage Company, a Moran booze depot at 2122 North Clark Street.

As another part of the plot, McGurn assigned Claude Maddox, leader of a small Capone-affiliated gang, to steal a car of the type used by Chicago detectives, install a police gong on the running board, and obtain some police uniforms. Next, McGurn hired some dependable trigger men, probably the following: Fred "Killer" Burke, a Detroit gunman wanted for murder, kidnapping, and bank robbery; Freddie Goetz, former manager of his fraternity house at the University of Illinois who had turned to professional murder; Albert Anselmi and John Scalise, whose efficient teamwork as torpedoes had earned them the nickname Homicide Squad; and Joseph Lolordo, a former World War I machine-gunner who had a grudge against Moran for killing his brother. Finally, McGurn recruited two members of the Detroit Purple Gang, Harry and Phil Keywell, to serve as scouts.

About 10:30 on the appointed Thursday morning the Keywell brothers, watching from a window in a boarding house across the street, saw a man fitting the description of Bugs Moran enter the front door of the S. M. C. Cartage Company. They made a telephone call and left. A few minutes later, the black car with police gong drew up to the curb, and four of the five men got out. Two wore police uniforms, and two wore heavy overcoats. The uniformed men entered first, guns drawn, and ordered those inside the garage to line up facing the wall. They did so, expecting no more than a routine arrest or a shakedown; then the two men in overcoats came in. Both carried Thompsons, which they emptied into the seven men lined up against the wall. Then the killers handed their guns to the "officers," raised their hands in the air, and marched back out to the car.

A neighbor who went to investigate the racket found the garage splattered with blood and littered with bodies—six Moran mobsters and one Dr. Reinhardt H. Schwimmer, a young optometrist who enjoyed hobnobbing with gangland characters and had stopped by that morning for a cup of coffee.

The only flaw in the entire operation was the absence of Moran himself. Another gang member resembled him closely enough that the Keywell brothers thought Moran had arrived, and summoned the executioners prematurely. Moran and two companions were approaching the garage on foot when they saw the phony detective car pull up in front. They turned around and left, or the death toll would have been ten.

Before 1929 Chicago had viewed gangland murders philosophically. The police tended to shrug them off as being in the public interest; and the public—at least much of it—regarded the beer wars as a sort of Underworld Series. A popular joke was that gangland killings "improved the breed," which they certainly did, and a violinist could consider himself fortunate if he got through an entire day without once hearing, "See you got your machine gun with you."

Out-of-towners felt cheated if their visit to Chicago did not include a glimpse of Scarface Al Capone (or a reasonable facsimile) rolling through downtown in something big and black enough to be a $25,000 armor-plated limousine with bulletproof windows.

But the St. Valentine's Day Massacre—seven men lined up and shot in the back—violated all the rules of fair play and sportsmanship, and civic reformers at last found more than token support. Reacting to the changed climate, the police launched an intensive investigation, beginning with a search for the submachine guns used by the killers.

Witnesses who had seen the killers leaving the garage recognized their weapons as the now-familiar Tommyguns, and the finding of seventy .45-caliber cartridge cases indicated, for the record, that one of the guns had had a fifty-round drum magazine and the other a twenty-round clip. Police began visiting the city's gun dealers, compiling a list of all Thompsons sold and their buyers. At the same time, the rumor began to spread that the killers were in fact Chicago policemen on a personal mission. Because of the police department's reputation for graft and corruption, the charge seemed plausible enough to warrant the ballistics testing of every police submachine gun in Cook County.

In their search for the murder weapons, police found that a gun dealer named Peter von Frantzius had sold six Thompsons to a man representing himself to be, appropriately, "F. Thompson," sporting-goods dealer in Kirkland, Illinois. Two more submachine guns had been sold to a man identifying himself as William McCarthy of the Indiana State Police. McCarthy turned out to be a fictitious name, and one of the guns sold to the supposed policeman soon turned up in a raid on the home of a Chicago beer-runner named Steve Oswald. When police located the residence of an F. Thompson, his wife explained that he was a very difficult man to see, as his job, soliciting students for a private school, kept him away from home most of the time. F. Thompson's aged parents, whom the police also questioned, were closer to the truth. They thought he sold cemetery plots.

The mystery of F. Thompson and his Thompson submachine guns was cleared up several months later, literally by accident. On December 14, 1929, two automobiles collided at a downtown intersection in St. Joseph, Michigan. The drivers began to argue, and city policeman Charles Skelly decided that any arguing should be done at the station. When he climbed on the running board of one of the cars, the driver pulled a pistol and shot him three times. The driver then damaged his own car further in his rush to get away, but quickly commandeered another at gunpoint and made good his escape.

The police hardly suspected any connection between the St. Valentine's Day Massacre and the Michigan shooting incident until they searched the fashionable Lake Shore Drive home of Frederic Dane, in whose name the abandoned car was registered. They found $319,000 in bank holdup loot and an arsenal that included two of the submachine guns purchased from von Frantzius by "F. Thompson." Frank Thompson, the police learned later, was little more than an underworld errand boy. But Frederic Dane turned out to be none other than Capone triggerman Fred "Killer" Burke.

In 1929 the science of ballistics was only beginning to find recognition and application in police work, and the St. Valentine's Day Massacre case provided the first conspicuous demonstration of its value as a law-enforcement tool. Colonel Calvin Goddard of New York, a pioneer in ballistics research and inventor of the comparison microscope, was asked to work on the case. Two years earlier Goddard had volunteered his services to the Massachusetts court reviewing the convictions of Sacco and Vanzetti. His findings, confirmed by a private inquiry in 1961, indicated Sacco guilty and Vanzetti innocent, but his evidence was "unofficial" and not acted upon.[15] In Chicago, however, Goddard appeared before a coroner's jury on December 23, 1929, and proved to the jury's satisfaction that the bullets taken from two of the Massacre victims had been fired from Fred Burke's submachine guns. Goddard also discovered that bullets from one of Burke's guns matched those found in the body of New York gangster Frankie Yale.

After months of investigation the police managed to reconstruct the Massacre in great detail, and the State of Illinois succeeded in getting murder indictments against several Capone mobsters. No convictions were ever obtained, of course, nor could Capone himself be touched. On February 14, 1929, he was vacationing at his Palm Island estate in Florida, and on the morning of the Massacre just happened to telephone the local chief of police to assure everyone that he was sitting at home and behaving himself.

The St. Valentine's Day Massacre was followed by a lull in the fighting, owing to pressure from the police and from underworld leaders elsewhere who regarded the slaughter as deplorable public relations. Another truce was arranged, and Capone himself vacationed for several months in a Philadelphia jail on a pistol-carrying conviction. When he returned to Chicago in March 1930, the underworld oiled up its machine guns.

On May 31, 1930, the Capone-affiliated gang of Frankie Lake and Terry Druggan unwisely shotgunned a member of the Joe Aiello gang, which had moved into the vacuum created by the St. Valentine's Day Massacre. Almost before the day was over the Aiellos retaliated. In the early morning hours of June 1, Aiello gunmen slipped up to the glassed-in porch of a Fox Lake resort hotel where a group of Druggans and Capones were sitting at a table. They poured a hundred machine-gun bullets through the windows, killing three and wounding two.

Capone struck back on July 29, killing Jack Zuta, the Aiello-Moran gang's bookkeeper. Zuta was hiding out at a Delafield, Wisconsin, resort hotel after an earlier attempt to kill him in Chicago. Not expecting visitors, he was preoccupied with feeding coins into a nickelodeon when five men walked onto the dance floor and motioned for everyone to stand back out of the way. As Zuta turned around, the lead gunman shot him with a pistol, then the others opened up with a Thompson and sawed-off shotguns. Zuta fell dead with fifteen wounds, and the killers calmly walked back out to their car.

Three months and several routine murders later, Joe Aiello himself was slain in one of the most thorough Tommygun murders of the Prohibition era. At 8:30 P.M. on the evening of October 23, 1930, he walked out of the front door of his luxurious apartment building at 205 Kolmar to get into a waiting cab. Suddenly a machine gun began firing from a nearby second-story window. Wounded,

Aiello tried to run back to his apartment building, just as a second machine gun and a sawed-off shotgun opened up from another second-story window. Aiello died with fifty-nine bullet wounds, probably gangland's record instance of overkill.

Aiello's was one of the most extravagant machine-gun murders of the Capone era, and also one of the last. Capone's violent tactics had wiped out much of his competition, but also had spawned a strong reform movement whose leaders labeled him "Public Enemy No. 1." He became the target of police harassment wherever he went, and even as he fled from city to city to escape his tormentors, federal tax agents were compiling the evidence that would send him to prison in 1932, for income tax evasion. While he was in Alcatraz his mind deteriorated from a syphilitic infection, probably a souvenir from his early days as a brothel bouncer. He was released from prison in 1939, a witless invalid, and died at his Palm Island estate in 1947.

The fall of Al Capone, the Depression, and finally Repeal, together with gangland's increasing concern for public relations, brought an end to Chicago's era of ostentatious violence. Local underworld leaders became more businesslike in their crimes and more subtle in their methods of dealing with one another. In New York, however, the trend seemed to be just the opposite.

While Chicago's gangsters had volleyed and thundered at one another and ridden through the city in kingly elegance, either in armored limousines or in $10,000 caskets, New York had maintained an air of righteous disapproval toward the Second City, sometimes sniffing with indignation, sometimes snickering at Chicago's plight. Gotham gangsters of the Twenties were gentlemen by comparison, or at least not so gross as to riddle their rivals in Times Square or herd them together for massacring. New York had her problem children, but they were not so conspicuous as the rowdies on the wrong side of the Hudson.

But with the 1930s, crime—at least the spectacular gangland murder—seemed to move east, and it came Chicago's turn to be smug. For when New York's gangsters started quarreling, they, too, adopted the Tommygun.[16]

Why the submachine gun reached New York so late is not too

clear, since New York already claimed as its own such underworld notables as Dutch Schultz, Legs Diamond, Vincent Coll, and Arnold Rothstein. Geography may have been a factor. The five-borough system with its clearly defined boundaries may have reduced the temptation to fudge across the other fellow's line. Also, New York gangsters seemed always to appreciate the importance of good public relations. Far easier to handle a police official who is not being pressured by the press and the public to "stop the killing in the streets!"

In addition, while the Chicagoans were still fighting beer wars, New York gangsters were busy developing the higher forms of "racketeering" that would become the mainstay of organized crime following Repeal. Men like Lucky Luciano, Joe Adonis, Lepke Buchalter, Bugsy Siegel, Meyer Lansky, Albert Anastasia, Longy Zwillman, and Frank Costello were a new breed of gangster-businessman and operated out of paneled, carpeted offices instead of flower shops and garages. Their criminal activities not only were organized, but "syndicated"—divided along operational rather than geographic lines, and governed more or less democratically. The top man in each racket possessed equal voting power in the Syndicate's decision-making process, and accepted majority rule as a wise alternative to gunfire.

The Syndicate came to power in the early 1930s, and it was during this period of transition—from Prohibition-era gangsterism to nationally organized racketeering—that murder by Tommygun made New York the country's new "crime capital."

As early as 1926 New York had already had one brush with machine guns. In Elizabeth, New Jersey, on the Thursday morning of October 14, seven armed robbers in two cars ambushed a small mail truck and got away with $161,000. They killed the driver with machine-gun fire and badly wounded his assistant, then broke into the truck with boltcutters while a Tommygunner sprayed the street to prevent any interference. A motorcycle patrolman escorting the truck was wounded and then struck by one of the bandit cars, and the wounded postal assistant was run over as the bandits drove away.

The crime outraged the East, and the New York papers instantly branded the robbers as Chicagoans, because they had used submachine guns. Worst of all, they had used them not against rival gangsters, but against an innocent mail-truck driver with a wife and family, and against the United States mails. On October 16, President Coolidge told reporters that the New Jersey mail robbery had been the principal topic of discussion at his cabinet meeting that week, and the next day the government announced that 2,500 Marines would be assigned to guard the mails. Ten days later the Post Office Department announced that it would furnish the Marine guards with 250 Thompson submachine guns.[17]

Frenzied police activity and public outrage continued well into December, the Thompson gun remaining the focal point of attention. On December 4, *Collier's* published a thrilling article titled "Machine Guns for Sale," the writer describing his great shock at discovering that a respectable New York City arms dealer would sell him a submachine gun. The next issue of *Collier's* pursued the subject further with an article titled "Machine Gun Madness," which dealt mainly with the McSwiggin murder in Chicago and the dismay it caused Marcellus Thompson.

New York made the most of the machine-gunning. It was thrilled, chilled, and outraged by such a crime, and placed the blame squarely on Chicago. New York would keep on blaming Chicago for future machine-gunnings, until they became too obviously the work of home-towners.

In 1928, when Frankie Yale was machine-gunned in Brooklyn, New York papers made it clear that the weapons, hence the killers, came from Chicago (as they probably did). Two years later, on March 7, 1930, another Tommygun killing took place in New Jersey that was clearly a local affair. The *New York Times* found a Chicago angle nonetheless. Its lead story on page one began, "The bold technique of the Chicago underworld was brought to within a block of City Hall and Police headquarters in Hoboken yesterday,"[18] and went on to describe the machine-gun slaying of Frankie Dunn, retired leader of some local beer-runners. Four armed men with two Thompsons caught Dunn in the lobby of an office building, then engaged police in a street battle while trying to escape. The police recovered one of the Thompsons and traced it to New York City patrolman Edward Beban. Beban admitted buying the gun in 1924 for an acquaintance, Thomas Coleman, who made his living supplying the Long Island rum fleet with provisions. Coleman told police that he had needed the gun because of certain hazards connected with his occupation. Police were unable to trace the man to whom he had later sold it.

As its own crime rate increased, New York only needled Chicago all the more. In 1928 and 1929 conditions were bad enough that New York Police Commissioner Grover Whalen launched a spectacular clean-up drive that featured special police strong-arm squads assigned to comb the city for "crooks and gunmen" and simply beat hell out of them. "There's a lot of law at the end of a night stick," Whalen declared.[19] When the first big drive failed utterly to bring results (police caught one pistol carrier, and he had a permit), the Police Commissioner proclaimed the drive a success: New York's criminals had got the message, obviously, and fled to Chicago. When informed of Whalen's remarks, Chicago Police Commissioner William F. Russell responded gamely, "We will send the New York crooks back to New York—in boxes, just as we did Frank Orlando, New York robber shot here the other day." Then, boasting that Chicago police had killed thirty "gangsters" and wounded fifty-four in the previous five months, he concluded, "We still have plenty of ammunition left and will be glad to use it for target practice on any sweepings from New York."[20]

In Detroit, home of the renowned Purple Gang, Police Commissioner Rutledge joined the contest by announcing a bounty of $10—a $10 gold piece, in fact—on each criminal killed by his men in the line of duty. "I notice there is undue police activity in New York and Chicago," he told reporters. "If Grover Whalen and Russell of Chicago are driving the crooks out of those cities, they won't come to Detroit if they know there is a price on their heads!"[21]

The efforts of the police commissioners to drive gunmen to one another's cities had no noticeable effect on homicide rates, but the spotlight still shined mostly on Chicago, and New York missed no chance to rub it in. On July 4, 1930, the *New York Times* published an especially sarcastic editorial in response to a Chicago gangland murder in which the killers escaped by means of a clever smoke-screen device attached to their car.

> [In Chicago] gang warfare continues to approach closer to the model of real war. Rifles and machine guns and bombs are an old story. . . . Now that fugitive taxicabs in Chicago emit smoke screens to facilitate their escape, we may look hopefully forward to Capone scout planes keeping an eye on Moran bomb planes; and in the Chicago River and in Lake Michigan miniature plane carriers and submarines; and in the Loop police tanks crushing their way through beer-runner barbed wire and over hijacker trenches; ultimately, a proposal from harassed civic sources for a conference between police and racketeers looking to a parity in guns, even if it be too much to hope for a reduction or complete disarmament.

This was New York's last smug editorial on Chicago crime. Several gang wars were brewing, and the continuing feud between Dutch Schultz and Vincent Coll came out in the open the following summer with the most publicized gangland crime in New York's history—the "Baby Massacre" of July 28, 1931.

Coll had been a teenage triggerman for Schultz in the Twenties, when Schultz was the more or less undisputed king of beer in the Bronx. The two parted enemies in January 1930, and Coll opened a rival beer-running business. The resulting dispute was Chicagoesque, and so was the shooting.

The feud's early victims, including Coll's brother, attracted little attention, being sandwiched in among the bodies that police routinely found in back alleys and vacant lots. Out of sight, out of mind, was the prevailing attitude toward gangland murders. But in July 1931 it was Coll's turn to shoot one of Schultz's men, and he unwisely chose to gun down a certain Joey Roa as Roa lounged in a chair in front of the Helmar Social Club at 208 East 107th Street. The day, July 28, was so far the hottest of the year—ninety-four degrees—and on the sidewalk next to Roa a fourteen-year-old boy was selling lemonade at a penny a glass. Nearby a baby lay sleeping in a wicker carriage. Other children were playing on the sidewalk and in the street.

Roa paid the children no attention, but he noticed instantly when an expensive touring car turned onto the street where expensive touring cars were seldom seen. He dove for the sidewalk, just as witnesses heard "sharp crackling like the explosions of firecrackers."[22] The car sped away, as did Roa. Behind him lay five wounded children, including the baby in the carriage, who later died.

Within hours newsboys were running through the streets shouting headlines like, "Babies machine-gunned by mad-dog killer!" Thus Mad Dog Coll earned the nickname he would carry, very shortly, to his grave.

No one was certain that a submachine gun had been used, but by this time almost every spectacular shooting was blamed on the Thompson. And with the "Baby Massacre," such things seemed to become a regular practice in New York. Only three days later and only three blocks away, two men were shot as they drove along East 103rd Street. Their car ran out of control and crashed, barely missing a number of children playing on the sidewalk. The same day, a clothing manufacturer was killed in front of his Brooklyn home by several shotgun blasts fired from a passing car. Three weeks later, a four-year-old girl, two policemen, and three hoodlums were killed, and twelve other persons wounded, in a wild twelve-mile bandit chase through the Bronx and upper Manhattan. One week after that, in Brooklyn, gunmen firing from a speeding auto missed their intended victim and hit an eighteen-year-old girl.

The Schultz-Coll war, after a time-out for Coll's trial and acquittal, continued into 1932, and that year drew to a bloody close with the Dutchman the winner. On the Monday night of February 1, Schultz men crashed a party at a house in the Bronx in search of "the Mick." Coll was absent, but the gunmen killed two of his men and a woman and wounded three other persons. This was the fourth or fifth unsuccessful attempt to get Coll, but a week later, on February 7, his luck finally ran out. While he was making a telephone call in the London Chemist drugstore at 314 West 23rd Street, a carload of Schultz gunmen pulled up in front. Two men waited outside the door of the shop. A third man, carrying a submachine gun, went in. After telling five horrified customers to "Keep cool, now,"[23] he calmly walked over to the telephone booth and sprayed it up and down with bullets.

Coll's sudden and rather novel demise in a bullet-riddled telephone booth captured the public's imagination and made phone booths nearly as ominous as violin cases. Also, it left Dutch Schultz one of the few surviving members of the old Prohibition dynasty of gangsters. Legs Diamond was dead. His celebrated ability to endure bullets was put to the test once too often as he slept in an Albany rooming house. The others—Arnold Rothstein, Vanny Higgins, Larry Fay, Waxey Gordon—were all out of the way, or soon would be, leaving Schultz to learn to live with the burgeoning Syndicate. This he refused to do, asserting his independence by vowing to kill reformer Thomas E. Dewey in violation of Syndicate orders that Dewey be spared for public-relations reasons. Syndicate gunmen, better known as Murder, Incorporated, shot Schultz to death in 1935 in the Palace Chop House in Newark with revolvers that were first reported as a Thompson submachine gun.[24]

6

The Depression Desperados

His pal was a very small woman
  Who smoked cigars, so they say;
The way she could use a machine gun
  Would turn a man's hair gray.
  —From "The Ending of Clyde and Bonnie," by W. F. Hill, 1934

By the end of the 1920s the submachine gun had established itself firmly in the public mind as a gangster weapon. So far as any newspaper reader could tell, only gangsters had submachine guns, and no doubt many people believed the widely circulated story that the Tommygun company gave free guns to some of the more prominent mobsters for the sake of advertising. In any case, no gangster could be a really first-rate gangster without one—at least in fiction, and to some extent in fact. In his autobiography, *The Stolen Years,* Roger Touhy tells of borrowing Thompsons and stacking them around his office strictly to impress some visiting Capone men that the suburban Touhy mob was as tough as any downtown.[1] In 1928 the British supposedly rejected the Thompson on the grounds that "we are gentlemen and refuse to descend to gangster levels."[2] Along with jazz, flappers, bathtub gin, speakeasies, one-way rides, and bulletproof limousines, the Tommygun became a familiar symbol of America's "roaring" Twenties.

Nor did the Thompson's reputation improve in the years to follow. If anything, it acquired even greater notoriety in the hands of armed robbers of the John Dillinger type, who came into vogue after 1930. No longer was it merely an instrument of gangland assassination. Now its bullets raked the public streets in small towns as well as cities, and America faced a new "tidal wave of criminality, continuous, limitless, torrential, which threatened to engulf and drown in its bloody waters a hundred million people." According to the Attorney General, anyway.[3]

The great "crime wave" of the 1930s may have been less a criminal phenomenon than a social and political one. Armed robbery was certainly nothing new, and if any tidal wave occurred, it was probably in crime statistics, which the Justice Department began compiling on a national basis for the first time in 1930. Nor was it anything new that the police could not catch "motorized bandits." State police agencies still were largely paper organizations, with no effective communications system or means of coordinating their efforts with local authorities or with the police of neighboring states. The famous bank robbers of the Thirties had better cars and better interstate roads, and they had submachine guns, but compared to some of their less-publicized predecessors of the Twenties they were short-lived, bungling amateurs who could not even meet expenses. In retrospect, they seem to have been less formidable than elusive. They acquired their enduring notoriety partly because the New Deal Justice Department needed some national villains in order to inaugurate its program of federalized crime control over strong states'-rights opposition, and partly because the Depression-weary public needed some new heroes, scapegoats, and vicarious adventure.

During the prosperous Twenties, when the businessman was king, and politics "just another racket," the successful gangster could be romanticized as a kind of robber baron who only gave the public what it wanted. "I make my money by supplying a public demand," said

Al Capone. "I call myself a businessman. When I sell liquor, it's bootlegging. When my patrons serve it on a silver tray on Lake Shore Drive, it's hospitality."[4] From a distance, at least, the gangster's life seemed excitingly dangerous and enormously profitable, and many a "poor working stiff" would have risked a few bullet wounds to rate the best tables in the best night clubs and to fill men's hearts with either fear or envy. The country was full of Reinhardt Schwimmers.

But with the Depression came a change in the American mood, and also a change in the type of criminal the public could secretly admire. After 1930 a poor working stiff considered himself lucky just to have a job, and no man in a breadline could find much glamour in a swaggering gangland fop. It was easier to smile at the exploits of a Robin Hood. If bootlegging was the respectable crime of the Twenties, Americans in the Depression could find a lot of good in the men who "only stole from the banks what the banks stole from the people."

Unlike the big-city gangsters—swarthy fellows with foreign names who bribed politicians and shot each other in the back—the Depression desperados were red-blooded all-American outlaws in the great Jesse James tradition who came from "good homes" and were "driven to crime" by one misfortune or another. They were underdogs in a nation of underdogs, boldly defying the vast (and largely discredited) forces of law and authority, making the police look foolish. Colorful, daring, often gallant, they robbed fat "banksters," led the cops a merry chase, and died with their boots on. At least so go the legends.

Also, they had machine guns, whose deadly reputation tended to rub off on any criminal who possessed such a weapon. In promoting the submachine gun, the Auto-Ordnance Corporation stressed that "Much of the usefulness of the Thompson Gun in the prison and law-enforcement field is derived from its psychological value. . . . The bolder convicts who would attempt escape or revolt in the face of slow firing, inaccurate hand weapons, are highly respectful of the withering barrage which can be laid down by submachine guns."[5] And so were police officers. Against a bunch of machine-gun-waving bandits, the small-town cop with a pistol was about as potent as an ancient Chinese soldier with stinkpot and gong.

But the Tommygun bandits of the Depression probably would never have earned their prominent place in American criminal history had their careers not coincided with the Roosevelt Administration's campaign to fight nationwide crime by means of national legislation.

The failure of state and local authorities to cope with organized interstate crime had become a national scandal by the end of the "Roaring Twenties," but neither state governments nor the Republican administration in Washington would concede any need for federal anticrime legislation. Law enforcement was, and always had been, the responsibility solely of the states. When public furor over the kidnap-murder of the Lindbergh baby forced Congress to enact a federal kidnaping law in 1932, President Hoover signed it reluctantly. His attorney general, William D. Mitchell, strongly opposed any further expansion of federal authority. He told Congress, "You are never going to correct the crime situation in this country by having Washington jump in."[6]

This attitude went out with the Republicans. When he took office in 1933, Roosevelt viewed lawlessness in the same New Deal light that he viewed unemployment: as a national problem requiring a federal solution. No longer was crime simply crime; now it was a "symptom of social disorder," and the new administration intended, in this area as in every other, to substitute "order for disorder."[7]

To plan a dramatic New Deal in national law enforcement, Roosevelt appointed as his attorney general Homer S. Cummings, an idealistic former state prosecutor who immediately set to work framing federal anticrime laws which would empower federal agents to swoop down on criminals interfering with "interstate commerce." Cummings' proposed laws received strong support from the President, and also from young Justice Department administrator J. Edgar Hoover. Since his appointment as director of the Bureau of Investigation in 1924, at the age of twenty-nine, Hoover had dreamed of building an American Scotland Yard, and had worked tirelessly at reforming and transforming his little-known bureau into a highly efficient investigative agency. The bureau had little authority, however. Its agents could neither make arrests nor even carry guns, except with local authorization, and had to content themselves mostly with catching "white slavers," interstate auto thieves, and violators of antitrust and bankruptcy laws. Cummings' federal crime legislation, if enacted, would vastly expand the bureau's jurisdiction, giving Hoover the national agency he considered essential in dealing with modern crime.

Together Hoover and Cummings set out, deliberately and altruistically, to convince a skeptical Congress that only federal laws and an effective federal enforcement agency could save America from underworld rule. In his *The FBI in Peace and War*, one of the more enduring of "FBI" books, Frederick L. Collins quotes Cummings' appraisal of the situation:

> State and local authorities were, in the very nature of things, virtually helpless. The more deeply we went into this problem, the more obvious it became that our first and perhaps our chief trouble was that crime was organized on a national basis, and that we were not. In short, what we needed was what we already had in the Federal Bureau of Investigation,[8] only we needed laws to free it from the shackles that had been placed upon it during the horseback era and send it forth to do its duty in a world of motor cars and machine guns. . . .

Collins goes on to describe Cummings and Hoover as two men who "almost alone among public officials in Washington, sensed the danger which threatened the American people [and] began preparing for the day when the people themselves would force their representatives in Congress so to implement the FBI that it could cope with this nationally organized crime—Cummings at his desk to draw the necessary laws, Hoover on the firing line to reconnoiter the enemy, to plan his campaign."[9]

Like Justice Department messiahs awaiting the day they would be called upon to lead their people out of underworld darkness, Cummings and Hoover were well aware that their people first would have to be educated as to their plight, lest the call never come. In short, they needed a national crime wave.

When gangbusters Cummings and Hoover set to work in the spring of 1933, the country had not a single big-name outlaw to its credit. Clyde Barrow and Bonnie Parker were acquiring a good deal of notoriety in Texas, but the only bandit to have attracted any national attention was Pretty Boy Floyd, a rowdy Oklahoma farmboy who had been robbing small-town banks with a submachine gun and a flair for the dramatic. Floyd was not the first armed robber to discover the value of the submachine gun, but he was the first of the Depression outlaws to make himself conspicuous with one, terrorizing Oklahoma like a one-man Dalton gang. Called Pretty Boy by hometowners amused at his pompadour haircut and foppish attire, Floyd became something of a hero to the Depression-stricken farmers of eastern Oklahoma by robbing unpopular banks and then throwing his money around. "Pretty Boy paid their mortgage and saved their little home," according to the popular Woody Guthrie ballad.[10] Floyd's generosity bought him good will, protection, and a measure of fame, as he probably intended it to do, and in December 1932 the *Literary Digest* published a short feature on his exploits, calling him "Oklahoma's 'Bandit King.' "

> He robs and laughs. Jeering the police, and even the Governor, he swoops down on a town, holds up a bank, and dashes away again by motor. In two years he has held up at least a score of banks. . . . But nobody has been able to bring him down. And this despite the tremendous chances he takes. That he is one of the luckiest bandits in criminal history is obvious from reading of his exploits.[11]

By the time he received this national recognition, however, the Bandit King had quit the bank robbing business and gone into deep hiding, and probably would have drifted back into obscurity but for some bad shooting in Kansas City a few months later. To raise some badly needed funds, Floyd hired himself out to rescue convicted bank robber Frank "Jelly" Nash from the guards taking him to Leavenworth by way of Kansas City, where he was to be transferred from a train to an automobile. On the morning of June 17, 1933, Floyd and two accomplices, armed with submachine guns, intercepted Nash and his guards in the Union Station parking lot. As the officers pushed Nash into the front seat of a waiting car, Floyd inexplicably opened fire with his Thompson, killing three policemen, a federal agent, and stupidly enough, the man he was supposed to rescue.

The Kansas City Massacre was the "machine-gun challenge to a nation" (as the *Literary Digest*[12] called it) for which the Justice Department had been waiting, and it came on the morning that the country's newspapers were bannering the abduction of wealthy St. Paul brewer William Hamm, Jr. This was the most sensational kidnaping since that of the Lindbergh baby the year before, and the Hamm kidnaping was overshadowed less than four weeks later by the kidnaping of Oklahoma City oilman Charles Urschel, whose abductors demanded an unprecedented $200,000 ransom.

Given two sensational kidnapings and a "massacre," Attorney General Cummings declared "war on crime" and dramatically unveiled his federal crime-control bills. "It is almost like a military engagement between the forces of law and order and the underworld army, heavily armed. . . . It is a campaign to wipe out the public enemy and it will proceed until it succeeds,"[13] Cummings declared after the Hamm kidnaping and Kansas City Massacre. After the Urschel kidnaping it was no longer "almost" a war. In a nationwide radio speech Cummings reported the country engaged in "real warfare which an armed underworld is waging upon organized society."[14] To fight such a war, of course, the Justice Department needed weapons. Congress quickly authorized the arming of the bureau's agents, and President Roosevelt obligingly instructed Cummings to study the advisability of transforming the bureau into a "superpolice force."[15]

The official excitement in Washington made good newspaper copy, and quickly shifted public interest from the big-city bootleggers to interstate bandits and kidnapers. Hoover, borrowing an idea from Chicago's reform movement, captured the public's imagination by establishing a hierarchy of "public enemies." The fact that most "public enemies" wielded submachine guns enhanced the Thompson's bloody reputation, and now the FBI had submachine guns too. With as much enthusiasm as alarm, Americans set aside some of their Depression worries and settled down with their newspapers to enjoy a national game of cops and robbers.

At the time Homer Cummings declared war on crime, the opposing armies were not very clearly defined. The public knew little of Hoover and his agents, who were not yet called G-men, or of the FBI, then known only as the Bureau of Investigation of the Department of Justice. To popularize the bureau and its men was one of Hoover's first tasks. The public not only had lost confidence in the law, he believed, but had developed an unhealthy admiration for the criminal. To reverse these attitudes, Hoover worked hard to build an image of his agents as lawmen of a new kind—federal, fearless, and incorruptible, heroic in the eyes of children and respected in the eyes of men. But to build such an image also meant, or at least resulted in, personalizing the Enemy into adversaries of exceptional challenge. With Pretty Boy Floyd yet to be identified as the killer of a federal agent in the Kansas City Massacre, the country had to make do for a time with Machine Gun Kelly.

George Kelly, an undistinguished bootlegger and small-time robber, participated in the kidnaping of Charles Urschel, supposedly at the urging of his ambitious wife, Kathryn. The newly enacted "Lindbergh law" brought the kidnaping under federal jurisdiction, and Urschel, released unharmed after payment of the ransom, enabled federal agents to break the case swiftly by remembering the exact time each day an airliner flew over the cabin where he had been held. Within a month the ransom had been recovered and all of the kidnapers arrested—except for Kathryn Kelly and George Kelly, who, by the third week of August 1933, somehow had become known as "Machine Gun" Kelly. The Kellys were the object of an exciting nationwide manhunt by Cummings and Hoover, and George Kelly's reputation as a desperate outlaw increased with every day federal agents failed to catch him. When he was finally captured in Memphis on September 26, without a shot, the *New York Times* reported that "Kelly and his gang of Southwestern desperados are regarded as the most dangerous ever encountered," and that the outlaw, according to the Justice Department, was a "desperate character, having served a number of prison terms [two, actually] for bootlegging, vagrancy, and other offenses."[16] After a sensational trial in Oklahoma City, Kathryn and Machine Gun Kelly went to Alcatraz for life, apparently without ever having fired a shot in anger.

How Kelly got his impressive nickname is not too clear. According to Hoover, Kathryn gave it to him to increase his prestige among her snobbish underworld friends. However, the name first appeared in newspapers only after Kelly had been tentatively identified as the kidnaper who walked into the Urschel home carrying a Tommygun. Federal agents captured a Thompson from another member of the gang and traced it to a Fort Worth pawnbroker, who identified the purchaser as Kathryn Kelly. This discovery, about the third week of August 1933, may have inspired Director Hoover to a little poetic license. In his book of famous FBI cases, *Persons in Hiding*, he wrote of Kathryn:

> . . . she rode into the country with him [George] while he practised to become an expert, finally reaching the point where he could knock a row of walnuts from the top of a fence and at a good distance. Meanwhile, Kathryn garnered the empty cartridges, to be kept for such times as she could hand them to friends, remarking:
>
> "Here's a souvenir I brought you. It's a cartridge fired by George's machine gun—Machine-Gun Kelly, you know."[17]

In solving the Urschel kidnaping, the Justice Department scored its first victory in the war on crime. The investigation had covered seventeen states, with arrests in five. Within ninety days the crime had been solved, the ransom recovered, and the kidnapers all sent to prison, clearly demonstrating the effectiveness of the federal government in dealing with interstate criminals. Justice Department press conferences and unprecedented newsreel coverage of federal courtroom trials, plus theatrical precautions against the prisoners' escape, insured that the public got the message.

The enormous publicity attending the capture and trial of George "Machine Gun" Kelly catapulted the "G-men" into national prominence. Kelly was a national figure, as were other desperados whose crimes seemed out of the ordinary. At the time, the two most extraordinary criminals by far were Clyde Barrow and Bonnie Parker, who had been terrorizing Texas and the Southwest since the spring of 1932. Unlike the comparatively gentle Kelly, who whiled away the hours playing cards with the man he kidnaped, the "Barrow gang" lived and robbed and killed in a thoroughly obnoxious fashion. Contrary to legend, they preyed mostly on gas stations and grocery stores, and seem to have had few of the redeeming qualities attributed to them later by writers and moviemakers. But they had miraculous luck in shooting their way out of police traps, exhibited desperate courage and loyalty to one another, and had an intimidating arsenal of weapons ranging from pistols and sawed-off shotguns to submachine guns and even Browning automatic rifles. Also, they were colorful. They took comic snapshots of each other, and Bonnie wrote epic doggerel on the romance of outlawry. Her final poem, which she gave to her mother, who later released it to the newspapers, was "The Story of Bonnie and Clyde":

> You've read the story of Jesse James—
> Of how he lived and died;
> If you're still in need
> Of something to read
> Here's the story of Bonnie and Clyde.
>
> Now Bonnie and Clyde are the Barrow gang.
> I'm sure you all have read
> How they rob and steal
> And those who squeal
> Are usually found dying or dead.
>
> There's lots of untruths to these write-ups;
> They're not so ruthless as that;
> Their nature is raw;
> They hate all the law—
> The stool pigeons, spotters, and rats.
>
> They call them cold-blooded killers;
> They say they are heartless and mean;
> But I say this with pride,
> That I once knew Clyde
> When he was honest and upright and clean.
>
> But the laws fooled around,
> Kept taking him down
> And locking him up in a cell

Till he said to me,
"I'll never be free,
So I'll meet a few of them in hell."

The road was so dimly lighted;
There were no highway signs to guide;
But they made up their minds
If all roads were blind,
They wouldn't give up till they died.

The road gets dimmer and dimmer;
Sometimes you can hardly see;
But it's fight, man to man,
And do all you can,
For they know they can never be free.

From heart-break some people have suffered;
From weariness some people have died;
But take it all in all,
Our troubles are small
Till we get like Bonnie and Clyde.

If a policeman is killed in Dallas,
And they have no clew or guide;
If they can't find a fiend,
They just wipe their slate clean
And hang it on Bonnie and Clyde.

There's two crimes committed in America
Not accredited to the Barrow mob;
They had no hand
In the kidnap demand,
Nor the Kansas City Depot job.

A newsboy once said to his buddy:
"I wish old Clyde would get jumped;
In these awful hard times
We'd make a few dimes
If five or six cops would get bumped."

The police haven't got the report yet,
But Clyde called me up today;
    He said, "Don't start any fights—
    We aren't working nights—
We're joining the NRA."

From Irving to West Dallas viaduct
Is known as the Great Divide,
    Where the women are kin,
    And the men are men,
And they won't "stool" on Bonnie and Clyde.

If they try to act like citizens
And rent them a nice little flat,
    About the third night
    They're invited to fight
By a sub-gun's rat-tat-tat.

They don't think they're too smart or desperate,
They know that the law always wins;
    They've been shot at before,
    But they do not ignore
That death is the wages of sin.

Some day they'll go down together;
They'll bury them side by side;
    To few it'll be grief—
    To the law a relief—
But it's death for Bonnie and Clyde.[18]

They went down together on May 23, 1934, riddled by police bullets fired from ambush on a country road between Sailes and Gibsland in western Louisiana. The pair were federal fugitives by virtue of having driven a stolen car from Texas into Oklahoma a year earlier, and they had been tracked to Louisiana by federal agents. But the opportunity for ambush came unexpectedly, at a time the federal agents could not be reached, and the honor of bringing down the outlaws went instead to Frank Hamer, a glamorous former Texas

Ranger captain assisting on the case. Commented Hamer later, "I hate to bust a cap on a woman, especially when she's sitting down."[19]

Of all the Depression outlaws, the one most renowned for style, elusiveness, and professional ability was John Dillinger. According to the more sentimental accounts, Dillinger, raised on an Indiana farm, was a good boy from a good home who fell in with a bad crowd. In 1924 he bungled his first holdup and was sent to prison, where he fell in with an even worse crowd. Embittered by an unusually harsh first-offense sentence of ten to twenty years, he joined with several veteran bank robbers in organizing a gang that would go into action as soon as he was released and could help them escape. Paroled in May 1933, Dillinger quickly pulled several small robberies to raise working capital, then smuggled pistols to his friends in prison. They succeeded in breaking out on September 26, the same day Machine Gun Kelly was arrested in Memphis. On October 12 they returned Dillinger's favor, their young benefactor having gotten himself captured in the meantime. With everyone finally present, the gang, led by Harry Pierpont, raided two Indiana police stations for submachine guns and began a bank-robbing spree that soon attracted national attention.

The gang's technique was polished, effective, and rather flamboyant. Numbering four or five, they would storm into a bank waving submachine guns, order everyone to lie down or crowd into a corner, and hurry out a few minutes later with loot and hostages. The event usually was enlivened with gunfire as the bandits kept police at bay or covered their retreat. On a few occasions their escape was hampered less by police than by crowds flocking to glimpse the spectacle of their very own bank being robbed by the notorious John Dillinger. For newspapers had played up Dillinger, rather than Pierpont, as the gang's leader.

Dillinger went to some effort to live up to his reputation as a bold, gallant outlaw. Inside a bank he remained composed and businesslike despite flying bullets, hysterical customers, and clanging alarms; in escaping, he treated hostages with good-humored courtesy and released them unharmed. A dozen or more people were killed during

Dillinger robberies and gun battles with police, but he himself is believed to have caused the death of only one, a bank guard who was shooting at him.

On January 25, 1934, the city police of Tucson, Arizona, captured Dillinger and the rest of the gang almost accidentally and without firing a shot. This made Dillinger the center of national attention, which he seemed to enjoy immensely. While several states squabbled bitterly over his extradition, Dillinger joked with reporters, complimented his captors, and chatted amiably with the crowds of people who were permitted to see him in the Tucson city jail. By the time he was transferred to a new "escapeproof" jail at Crown Point, Indiana—where he posed for pictures with his arm around the prosecutor who had vowed to send him to the electric chair—Dillinger was something of a national pet tiger. Then, on March 3, he accomplished the impossible. Using what he later claimed was a wooden pistol,[20] he locked up a score of jailers, seized two machine guns, a car, and two hostages, and motored leisurely out of town without attracting the attention of either the National Guardsmen or the armed vigilantes guarding the front of the jail.

Dillinger's incredible escape astonished and somewhat amused the entire country, and seemed to prove the Justice Department's contention that local and state authorities alone could not handle big-league criminals. Obviously, this was a job for the FBI, and Dillinger finally gave Hoover a chance to prove it. In making his getaway, the outlaw broke his first federal law by thoughtlessly driving a stolen car across a state line.

The G-men moved swiftly but not too effectively. On March 31, agents stumbled onto Dillinger in a St. Paul apartment house. He drove them back with machine-gun fire, grabbed his girl friend, and escaped down an unguarded back stairs to a car. Its second chance the FBI botched sensationally. On April 22, agents closed in too hurriedly on Dillinger and his new gang at a Wisconsin resort hotel—called Little Bohemia—and opened a roaring gun battle by mistakenly shooting three innocent customers who walked out to their car at the wrong moment and either failed to hear or refused to heed an agent's order to surrender. One of the customers died; later that night an FBI agent was killed and two officers with him were wounded. But Dillinger and his gang escaped again.

The Little Bohemia fiasco made news around the world. It also made Dillinger seem nearly invincible, an image that law enforcement agencies tended to foster by way of accounting for their inability to bring the outlaw down. Attorney General Cummings demanded airplanes and armored cars, and state governors called out the National Guard. Police, guardsmen, and possemen numbering some five thousand opened the country's greatest manhunt.

To the public, such extraordinary measures made Dillinger an extraordinary outlaw. He was made so officially when J. Edgar Hoover proclaimed him "Public Enemy No. 1," pulling out the old publicity trick that Chicago had used to help turn public sentiment against Al Capone. When he finally died on July 22, 1934—betrayed by a "woman in red" and shot from behind by federal agents as he left Chicago's Biograph Theater—the Dillinger legend was complete.

Once Dillinger had fallen, the rather melodramatic title of Public Enemy No. 1 passed to Pretty Boy Floyd, whom the FBI had since identified as one of the machine-gunners in the Kansas City Massacre. On October 22, 1934, after one gun battle and two weeks of manhunts, Federal Bureau of Investigation agents with Thompsons cornered Floyd on an Ohio farm and shot him fourteen times as he tried to escape across a field.

Which elevated Baby Face Nelson to the top of the FBI's Public Enemy list. Nelson, whose real name was Lester Gillis, was a minor member of the gang that Dillinger had put together following his escape from Crown Point. He distinguished himself mainly by killing a federal agent during the gang's escape from the Little Bohemia resort hotel in April. Unlike Dillinger, the gentleman bandit, or Floyd, the Oklahoma Robin Hood, Nelson was a colorless, hotheaded, psychopathic killer whom Dillinger had had to lecture from time to time about unnecessary shooting. The killing of a federal agent, however, made him a criminal celebrity. And the killing of two more in his last fight with the law guaranteed him a prominent place in the annals of villainy. On November 27, 1934, near Barrington, Illinois, G-men spotted Nelson's car and managed to disable it in a running gun battle. Two agents, Herman Hollis and Sam Cowley, armed with a sawed-off shotgun and a submachine gun, stopped their own car not far from Nelson's and began firing. According to witnesses, Nelson recklessly charged the agents, firing his Thompson from the hip, killing both of them. He himself was hit seventeen times, but still managed to escape—in the agents' car—with the help of his wife and an accomplice. The next morning police found his body abandoned in a ditch some twenty miles away.

The last of the Dillinger-era desperados to acquire notoriety in a spectacular machine-gun battle was "Ma" Barker. A dowdy old woman of Ozark hillbilly stock, Arizona Clark Barker had managed the criminal careers of her four sons and a young man named Alvin Karpis in everything from petty thievery to armed robbery and murder throughout the 1920s. In July 1932 they robbed a bank in Concordia, Kansas, of $240,000, and in December of that year they killed two police officers and a civilian with submachine guns in the course of robbing a bank in Minneapolis. The "Barker-Karpis gang" did not attract any national attention, however, until it turned to kidnaping and its members became federal fugitives. They abducted William A. Hamm, Jr., in June 1933 and banker Edward G. Bremer in January 1934, both in St. Paul, collecting ransoms totaling $300,000. After many months of thorough and highly sophisticated investigation, the FBI solved both kidnapings, capturing or killing all of the principals and developing solid evidence that convicted dozens of accomplices in many different states. The highlight of the case was the killing of Ma Barker and her son Fred in January 1935. Agents had tracked the two all over the country, finally locating them in a two-story frame house near the little town of Oklawaha, Florida. Early on the morning of January 16, agents surrounded the house and ordered the Barkers to surrender. They were answered by a burst of machine-gun fire from a second-story window. The battle lasted more than an hour and brought death to Freddie and Ma Barker, but also instant fame. Ma Barker herself apparently had no criminal record, and was unknown to the public until newspapers front-paged the story of her sensational last stand. Soon her career in crime thrilled the readers of magazines and books. According to one account, "They had died in the old Oklahoma tradition, empty guns in their hands. With one final filial gesture Fred Barker had given his old mother the gun which carried a magazine holding one hundred rounds. The good son kept the 50-shot machine-gun for himself."[21]

In context this account was sarcastic, but like most crime writing of the period, it still tended to glamorize the outlaws. It helped to create and propagate the basic elements of the Ma Barker legend: that she was a stouthearted woman outlaw in the Belle Starr tradition who, with her loyal son, chose to die fighting in a gallant but hopeless battle with the superior forces of the law. Even J. Edgar Hoover's abusive description of her, in an *American Magazine* article written in collaboration with Courtney Ryley Cooper, a particularly melodramatic writer of that period, tended to enhance Ma Barker's image as a colorful archcriminal.

> The eyes of Arizona Clark Barker, by the way, always fascinated me. They were queerly direct, penetrating, hot with some strangely smouldering flame, yet withal as hypnotically cold as the muzzle of a gun.[22]

Likewise, Hoover's earlier tribute to the agents killed by Baby Face Nelson had had the effect, certainly unintended, of transforming Nelson from a common murderer into an outstanding "Public Enemy":

> They were soldiers, fighting side by side with other brave and honorable soldiers of the Bureau against an enemy of the people. . . . There was no martial roll of drums to buoy up their spirits as they went to face their death, no crash of cymbals, no blare of bugle or trumpets—only the vicious rattle of a machine gun.[23]

By the end of 1934 Attorney General Cummings, with the help of President Roosevelt, J. Edgar Hoover, and assorted Public Enemies, had obtained a series of new and far-reaching federal anticrime laws that made the state line a symbol of danger instead of safety to an outlaw on the run. This was the beginning, too, of the modern

FBI, with its national fingerprint file and crime laboratory serving state and local law-enforcement agencies. And it was the end of glamorous outlawry. Millions of "Junior G-Men" soon put their toy Tommyguns to work "on the side of law and order," the side on which Auto-Ordnance had always intended that the real guns be used.

From the exciting contest between the G-men and the Public Enemies of the Thirties, Americans learned that Crime Does Not Pay. They also learned that it could be a short cut to fame and immortality; and that the mark of a really important criminal was a Thompson submachine gun. But if the sensationalized exploits of the John Dillingers and Pretty Boy Floyds and Baby Face Nelsons fascinated the public, they thoroughly dismayed those responsible for maintaining law and order. Particularly alarming was the fact that such criminals had machine guns. In 1934 the *New York Times* editorialized:

> It is interesting to note the different role which it [the machine gun] has come to play as a military weapon and as a gunman's weapon. The outstanding development of the World War was said to be the effect of the machine-gun in enormously strengthening the defense. Captain Liddell Hart is not alone in believing that the rapid fire gun has shattered the old theory of victory on the side of the heavy battalions. One soldier with such a weapon may under favorable circumstances hold back a thousand men.
>
> Among the criminals, on the other hand, the machine-gun has greatly augmented the power of the offensive. A couple of desperate men may dominate a crowd. A familiar and sinister item in our news is the speeding bandit motor car with its machine-gun "spraying" the landscape. If there was a need for laws dealing with the sale of deadly weapons in the comparatively simple days of the pistol, the problem becomes imperative when a single thug can equip himself with the firing power of a company.[24]

The firepower of a submachine gun, together with its murderous reputation, did in fact give an enormous advantage to the criminal armed with such a weapon. In 1922 John Thompson assured police officials that "any man will think twice before going up against one."[25] In 1926 a *Collier's* writer declared, "This Thompson sub-machine gun is nothing less than a diabolical engine of death. . . . With it one bandit can stand off a whole platoon of policemen."[26] In 1934 an article in the magazine *Today* reviewed John Dillinger's adventures with Tommyguns:

> When Bandit John Dillinger replied with a machine gun salvo to the Federal police attack upon his hide-out in the Little Bohemia resort in the Wisconsin woods in April, he was firing back with police artillery. . . . Dillinger and his gang were firing submachine guns obtained either by raid on police arsenals or through other channels which had legal entrances and illegal exits. In the few weeks since his ludicrous "toy-pistol" escape from the Crown Point, Indiana, jail, March 3, he had collected a dozen sub-machine guns and automatic rifles. . . .
>
> Onlookers say that the Dillinger crew cleared the streets of Mason City, Iowa, with fire from seven or eight sub-machine guns when it robbed a bank there of $50,000 on March 13, only ten days after the leader's jail break. . . .
>
> He lost some of the weapons presently to his pursuers. Two or more sub-machine guns were left when he fled, wounded, from a St. Paul apartment. His replenishing visit to the Warsaw, Indiana, jail followed. Several sub-machine guns were abandoned in the escape from Little Bohemia. Yet the scattering bandits carried one sub-machine gun each, or better, in the new flight.
>
> When machine guns had mere gang use, the volleys killed other gangsters. The public, and sometimes the police, were at cynical ease. What matter if the outlaws did get artillery! The more they blasted each other, the better. But the guns are now aimed at police and citizenry.[27]

Once the bootlegging gangs focused attention on the submachine gun, many states outlawed the private ownership or possession of fully automatic weapons. The Depression bandits caused the submachine gun to be outlawed nationally. Among the federal anti-crime bills presented to Congress in 1934 was a proposal to regulate the sale of firearms. After a gun battle of a different sort—a battle "to disarm the gunmen"—the Justice Department managed to secure the passage of the National Firearms Act, which did much to end the Tommygun's spectacular criminal career.

When gangland machine guns first made headlines the question asked by public officials and editorial writers was, How are criminals able to arm themselves with such terrible weapons? The answer was simple: they bought them from their friendly neighborhood gun dealer, who either cared not or dared not to ask too many questions. Sometimes they ordered them in wholesale lots direct from Auto-Ordnance or through one of its regional distributors, using nothing more than letterhead stationery for some fictitious export or sporting-goods concern. Because police and sheriff's departments, banks, private protective agencies, armored car companies, and the like frequently purchased their equipment through local dealers, Auto-Ordnance distributed the Tommygun through any armament or hardware firm interested in doing the business. Most of the distributors were wholesale firms; some were sales agencies. They took orders from other wholesalers and retail dealers in their area. Consequently the gun went to almost any purchaser who had a convincing letterhead, and thence to anyone who had cash and (if the retailer had any scruples at all) an honest face.

The scruples of some dealers were none too strict. In Chicago, gun dealers Alex Korecek and Peter von Frantzius not only sold machine guns to any seedy character who had the cash, but obligingly ground off the serial numbers at the customer's request. In Philadelphia, Edward Goldberg, a dealer in military equipment accused of selling Thompsons to local gangsters, told a grand jury that he didn't know who bought them and didn't consider it any of his business so long as he got the money.[28] In New York, a magazine writer[29] doing an exposé ("Machine Guns for Sale") for *Collier's* ordered a Thompson by mail from a local dealer in police equipment by way of proving his point. The dealer took the order, but sent him a city detective instead. The detective gladly explained to the writer how dangerously simple it was for anyone to buy machine guns.

The underworld's most brazen purchase of submachine guns was made directly from the Auto-Ordnance Corporation. In January 1928 the Auto-Ordnance main office in New York received a letter from the "Mexamerica Company" of Chicago asking for the jobber's price on a quantity of Thompsons. Ever alert, Auto-Ordnance tried to look up the Mexamerica Company in a trade magazine's list of sporting-goods jobbers. When it could find no listing for such a firm, Auto-Ordnance requested more information. Mexamerica replied that of course it was not listed, being a new company, strictly wholesale and export, whose customers were mostly revolutionists in Mexico and South America. This satisfied Auto-Ordnance. Prices were quoted, a check was received by return mail, and the order was shipped the same day by the Auto-Ordnance office in Hartford. Then the check bounced, and the Chicago address of the Mexamerica Company turned out to be that of a little Italian shoemaker who had happily accepted $10 from a stranger to hang a "Mexamerica" sign in his window and take delivery of a crate. Five months later, two of the guns turned up in a $133,000 mail-train robbery in the Chicago suburbs.[30]

At the time Assistant State's Attorney William McSwiggin was murdered, in 1926, Marcellus Thompson estimated that some twelve submachine guns had fallen into criminal hands in the Chicago area. The Auto-Ordnance sales agent in Chicago guessed the number to be as high as forty.[31] The National Crime Commission, which met in Chicago later that year, declared that at least eight hundred submachine guns had reached criminals in all parts of the country.[32] In 1930 New York City police conducted an investigation into machine gun sales and concluded that no less than 12 per cent were to fictitious concerns or individuals—presumably criminals. The police list was compiled from the records of "one concern in this city"—presumably Auto-Ordnance:[33]

Law enforcement agencies	297
New York City concerns	91
Prisons	84
Fictitious buyers	83
Mexico	61
Individuals	47
Gun dealers	39

U.S. Marine Corps	33
Business and industrial firms	32
Banks	13
Armored car companies	5
Total	785

The 12 per cent does not quite work out mathematically unless the 91 guns sold to New York City concerns are subtracted from the total. These guns may have been duplicated in the other categories. But the fact that 83 of the guns went to fictitious buyers is impressive nonetheless.

Despite the apparent availability of Tommyguns at standard cost through even the simplest of ploys, the Thompson supposedly commanded exorbitant prices in the underworld. In his 1926 *Collier's* interview, Marcellus Thompson said, "I know that there are gangs of crooks in various places who have offered as high as $1,000 for one of these guns."[34] Dealers in Chicago and Pittsburgh reportedly received this price, and a policeman in Cicero is supposed to have ordered several Thompsons through his department and sold them to gangsters for $2,000 apiece.[35] Herbert Corey, in *Farewell, Mr. Gangster!*, published in 1936, estimated Thompsons to be worth $1,500 on the underworld market, and Colt Monitors (a commercial version of the BAR) worth as much as $5,000.[36]

After state laws began to tighten, and during the police crackdowns on dealers that customarily followed any sensational shooting, criminals found it easier and far more economical to obtain machine guns by theft. National Guard armories were their favorite target. During one twenty-six-month period in the early Thirties armories lost 1,557 .45-caliber pistols, 92 .30-caliber rifles, 20 Thompsons and BARs, and 50 miscellaneous guns.[37] John Dillinger specialized in daring daylight raids on police stations, although he, too, bought guns—some of them custom made—from underworld dealers. In two separate gun battles with police, Dillinger left behind cleverly designed machine pistols—Colt .45 automatics modified for full-automatic fire and equipped with twenty-round clips, muzzle brakes, and Thompson vertical foregrips. These guns were eventually traced to H. S. Lebman, a dealer in San Antonio, Texas, who admitted making them and selling them to a man he identified from FBI pictures as Baby Face Nelson. When arrested in April 1934, Lebman had one Thompson in his shop and another machine pistol under construction on his worktable.[38]

One of the submachine guns left behind by Dillinger on another occasion was traced to an El Paso dealer, the Munsen-Dunnegan-Ryan Company, which had sold fifteen Thompsons to the local Mexican consul general in 1929. Since then, investigators learned, every one had been "lost or stolen."[39] When the Birger-Shelton bootleg war broke out in southern Illinois around 1926, both sides had friendly sheriffs and deputies who bought submachine guns through regular police channels and then conveniently mislaid them. In one town, the chief of police bought two Thompsons on behalf of a local roadhouse keeper. When questioned later by state authorities, the roadhouse keeper explained (according to the Auto-Ordnance salesman who helped the police trace the weapons) that he had taken the guns to Idaho to hunt, "lions, tigers, and bears." But while on his way to the big hunt he had left his car parked on the side of the road with the guns inside, and alas! some rascal had stolen them both.[40]

Local, state, and federal officials, special fact-finding commissions, magazine and newspaper editorial writers, even sportsmen's groups, all agreed that the national gun situation was chaotic and intolerable. No one, however, could agree on a solution. Attorney General Homer Cummings, predictably, wanted a comprehensive federal law that would cover every type of firearm, discourage or prevent mail-order sales, and closely regulate the sale, possession, and interstate transportation of guns. States'-rights advocates argued that uniform state laws would accomplish the same degree of firearms control without the further expansion of federal police powers. The country's "gun lobby," chiefly the National Rifle Association, saw a sportmen's Armageddon in the often vague and conflicting language of the many bills written by legislators having no understanding of firearms types and technology. (Some, for example, would have defined "automatic" according to magazine capacity, or classified firearms according to their caliber.) Appreciating the power of the antigun forces, the NRA conceded the need to regulate such "gangster" weapons as the submachine gun, but strongly opposed bills that would register or tax handguns, rifles, and shotguns. The association regarded such laws as unenforceable, except against the citizen who was law abiding anyway, and another step in the direction of gun confiscation.

Said Homer Cummings to the Daughters of the American Revolution, "There are more people in the underworld carrying deadly weapons than there are in the Army and Navy of the United States!"[41] Said the NRA to its members, "Every rifle and shot gun owner in the country will [with passage of the Cummings bill] find himself paying a special revenue tax and having himself fingerprinted and photographed for the 'rogues gallery' every time he buys or sells a gun of any description."[42] The states'-rights forces were divided as to the scope of any gun laws, but were unanimous in their opposition to any law that was federal. The Chairman of the House Judiciary Committee, Hatton Sumners of Texas, managed to bottle up Cummings' firearms and other bills because they "did violence" to states' rights. Then, on April 22, 1934, while the bills were stuck in committee, John Dillinger and his gang machine-gunned their way out of the FBI trap at Little Bohemia, leaving two dead and four wounded. The next day President Roosevelt called Chairman Sumners to an emergency White House conference. When Sumners came out he told newsmen that "A wrath like that which kindled the frontier when vigilantes cleaned out the gunmen is sweeping America," and that he would therefore ignore his personal feelings and rule for passage.[43]

A new federal gun control bill—the National Firearms Act—soon passed both houses and was signed into law in June 1934. It covered, however, only machine guns, sawed-off shotguns, sawed-off rifles, silencers, and concealable firearms *except* pistols and revolvers. To the regret of the Justice Department and other firearms foes, the Washington "gun lobby" succeeded in "emasculating" that part of the bill covering conventional handguns. Nor did the National

Firearms Act outlaw, strictly speaking, the private purchase or ownership of machine guns or the other weapons covered. Instead, it imposed a prohibitively high tax of $200 on the unlicensed making or transfer of such items, under the federal government's powers of taxation. In the case of "gangster" weapons, Cummings recognized, the power to tax was the power to destroy.

As a criminal's weapon, the Tommygun was an unqualified success. As a police weapon, it was such a flop that many law-enforcement officials wished sincerely that it had never come off the drawing board. When Auto-Ordnance sales manager J. E. T. Sturm went to the Chicago police to volunteer his help in the St. Valentine's Day Massacre case in 1929, Captain John Stege practically ran him out of the office. Said Stege, "I told him the only help he could give was to go back and close the gun factory."[44] It was Stege's opinion that not even the police should be armed with machine guns, and his opinion was shared to some extent by many other lawmen in the country. In 1933 the Senate opened an investigation into "racketeering" at which the topic of machine guns figured prominently. In their testimony before the committee, police officials almost unanimously denounced the Thompson as generally useless for police purposes. Chicago Police Commissioner Joseph P. Allman told the committee:

> I think the machine gun should be manufactured only by the Federal Government. I do not think any police department has any legitimate use for the machine gun. We have them in our ordnance department, but only as a matter of being as well prepared as the hoodlum.[45]

Colorado Springs Police Chief H. D. Harper rejected the submachine gun in favor of the shotgun:

> We have machine guns in our department. We also have sawed-off shotguns of the best type. . . . I know a good sawed-off shotgun is of a thousand times more value to your police department than a machine gun is. It is not possible for a police officer to open a machine gun up on a crowded street . . . because you are going to kill possibly ten innocent people to one criminal.[46]

The only real support for the submachine gun as a police weapon came from ballistics expert Calvin Goddard, and his support was qualified. Specifying that the gun be handled only by an expert, he pointed out that the Thompson:

> . . . has a bullet of low velocity and which shoots with considerable accuracy for a couple of hundred yards, but which does not endanger the community as the high-powered rifle does and which, at the same time, is infinitely more effective than a sawed-off shotgun. You cannot stop an automobile a hundred yards ahead of you with a shotgun, but you can do it with a machine gun every time.[47]

This was one of the Auto-Ordnance Corporation's main selling points, and it was a valid one. But seldom, if ever, was the town Tommygun handy on the rare occasion when it might be used safely and effectively against a worthwhile lawbreaker. The larger cities bought Thompsons in the 1920s mostly because the bootleggers had them, and by the early Thirties the guns could be found in the arsenals of even the smallest police and sheriffs' departments of the Midwest and Southwest, a provision for the big day that Dillinger or Pretty Boy or Clyde and Bonnie might come through town. But in cities and towns alike, the Thompson found far more use as a showpiece than as a weapon.

In December 1926, Chicago police purchased thirty-four submachine guns "for trial," only to return most of them after a few months on the grounds that they were not useful.[48] Six years later, Chief of Detectives William Schoemaker told the Senate rackets committee: "In my experience with [the submachine gun] we have never used it, that is, to fire it, but we have carried them. They are intimidating to the gangsters when you have a machine gun with you, but we have never fired one that I know of."[49] A few months later, in December 1933, the Chicago police finally used their Thompsons (or at least carried them) in a raid on an apartment where Dillinger was believed to be hiding out. The three men shot and killed had no connection with the Dillinger gang, but fortunately they turned out to be wanted criminals anyway.

New York City had armed special "riot squads" with prototype

Thompsons as early as 1920, and in 1925 Police Commissioner Richard Enright equipped the city's new emergency wagons with six Thompsons apiece. But the guns were not used in actual police work until May 7, 1931, when Francis "Two-Gun" Crowley engaged police in a sensational two-hour battle on Manhattan's upper West Side.[50] Crowley, wanted for murder and armed robbery, had been cornered on the fifth floor of an apartment building at 303 West 90th Street with his sixteen-year-old girl friend, Helen Walsh, and an accomplice, "Big Rudolph" Duringer. When he discovered the police were closing in, Crowley began shooting. The police returned the fire from the street and from nearby buildings, wounding him four times before finally subduing him with tear gas. Submachine guns were brought into the fight, but did not prove very useful. The Thompson was not capable of pin-point accuracy, and to turn one on full-automatic might have showered ricochet bullets into the thousands of spectators watching the battle from the streets and from surrounding apartment windows.

Federal agents seem to have had more opportunities, or less reluctance, to use submachine guns, but the guns were not always used to the agents' credit. In the attack on Dillinger at the Little Bohemia lodge, agents mistakenly shot three innocent customers. After the capture of bank-robber Harry Brunette in 1936, the FBI was strongly criticized for the way it laid siege to the apartment on New York's West 102nd Street where he was hiding. Although Brunette, unlike Crowley, was not putting up much of a fight, agents riddled the place, endangering not only the other tenants of the building but also firemen who had been called to fight a blaze started by a tear-gas grenade. *Newsweek*'s unflattering account included the following:

> Amid the hubbub, a flustered G-man poked a submachine gun at a husky fireman. "Dammit, can't you read?" growled the fireman, pointing at his helmet. "If you don't take that gun out of my stomach, I'll bash your head in." For thirty-five minutes, the shooting continued. Then a lull. "Give up, or we'll shoot," shouted a G-man—as if they had been throwing spitballs up to then.[51]

7

Rum, Riots, and Revolutions

You could not run a coal company without machine guns.
—*Attributed to industrialist Richard B. Mellon*

Despite its initial publicity and later notoriety, the Thompson submachine gun was a failure from the start. In 1921 Auto-Ordnance received orders for well over a thousand guns. But nearly half of these went to Irish revolutionaries and ended up in a Customs Bureau warehouse. Most of the others went to large industrial and mining firms or local police agencies for dealing with "Bolshevists" masquerading as union organizers and striking workmen. Belgium purchased a hundred for policing the Congo, and Panama and Mexico each bought about thirty, but the only government or military orders from other countries, including the United States, were for small sample lots of two or three for testing purposes. On February 17, 1923, Theodore Eickhoff calculated that the company had spent $1,039,887 for submachine guns, stocks, Lyman sights, spare parts, and manufacturing equipment, plus $17,185 for Autorifles, which did not include development, promotion, and normal operating expenses since 1916.[1] When the Model 1923 Thompson proved an instant flop, costing additional thousands, the only thing that kept Auto-Ordnance in business at all was the patience of Thomas Fortune Ryan.

After 1921, sales declined steadily, as did the company's prospects for the future. In 1922 Oscar Payne left Auto-Ordnance to design screw manufacturing machinery for the Reed & Prince Company of Worcester, Massachusetts, and in 1924 Theodore Eickhoff resigned to join the staff of the Trundle Engineering Company of Cleveland. About 1925 General Thompson, in his middle sixties, also retired from active work in the company, leaving business affairs largely in the hands of his son and George Goll, who had replaced Eickhoff as chief engineer.

Although deep in the red, the company theoretically had assets of nearly $3,000,000 in its inventory of Tommyguns, and Marcellus Thompson never ceased pestering the military services with new ideas and sales talks. During the second half of the Twenties his efforts slowly began to pay off, thanks to rumrunners and Nicaraguan rebels.

By 1925 the "rum war" along the eastern seaboard was no longer a gentlemanly contest of wits and speed between smugglers and the Coast Guard. After several routine chases were marred by unsportsmanlike gunfire, the Coast Guard began arming its seventy-five-foot patrol boats and larger craft with submachine guns for use by boarding parties. During the Prohibition years the Coast Guard purchased about five hundred Thompsons. After Repeal the gun remained in service as standard Coast Guard issue: one per vessel of more than seventy-four complement, one per air detachment, and three per air station.[2]

The Coast Guard employed the Tommygun more as a police than a military weapon, and then rarely had occasion to actually shoot it at anyone, legends notwithstanding. But the submachine gun's official use by an official government service constituted a minor breakthrough for Auto-Ordnance nonetheless. At last the Tommygun had the endorsement of the military, not just strike breakers and Al Capone.

After 1926 the Marines gave the Thompson further respectability. Like the Army, the Marine Corps had buzzed with excitement at news of the new weapon in 1920, praised its performance in demonstrations, and predicted its success as a military small arm. In 1921 the Marine Corps even worked up an elaborate recruiting poster showing ". . . up-to-the-minute views of Marines operating the Stokes mortar, Thompson submachine gun and the most modern weapons of the day." But nothing much came of Marine enthusiasm either, and as late as 1925 a board of officers with the Second Regiment Marines in Port-au-Prince, Haiti, tested the gun extensively and recommended that it not be adopted as a Marine infantry weapon.[3]

Then, in October 1926, "machine-gun bandits" attacked the mail truck in Elizabeth, New Jersey. The Post Office Department not only called out the Marines to guard the mails, but issued them 250 Thompson submachine guns—just like the kind the outlaws had used.

By the time the mail-robbery crisis had passed, the Marines were embroiled in new troubles in China and Nicaragua, and deeply grateful to the Post Office Department for gifting the Corps with what soon proved to be extremely handy weapons. Early in 1927 Chiang Kai-shek, leader of the Chinese Nationalist party, the Kuomintang, purged the party of its Communist faction (led by Mao Tse-tung), plunging the country into civil war that threatened foreign lives and property. In February of that year the United States landed a provisional battalion of some five hundred Marines in Shanghai. By fall 1927 several thousand Third Brigade Marines under Brigadier General Smedley Butler were stationed in Chinese port cities, mainly Shanghai and Tientsin, armed with 182 Post Office-issue Thompsons.[4]

Previously, a revolt and counterrevolt had occurred in Nicaragua, resulting in the landing of Marines in that country. In the fall of 1925 Nicaraguan Conservative leader Emiliano Chamorro overthrew the popularly elected Liberal government and installed himself as virtual dictator. In May 1926 a Liberal counterrevolt, together with the refusal of the United States to recognize the Chamorro regime, compelled the Nicaraguan congress to elect Adolfo Díaz as a compromise president. The Díaz government quickly received United

States recognition, plus several thousand Marines sent at Díaz's request to help suppress the Liberal forces. In addition, President Coolidge sent Henry L. Stimson to mediate. Stimson managed to persuade both the Conservatives and the Liberals to turn in their weapons and establish a Marine-trained, nonpartisan national guard to maintain order until free popular elections could be held, under Marine supervision, in 1928. One of the important Liberal military commanders refused, however, to enter into this arrangement. General Augusto Sandino led his rebel forces into the jungles where, despite a Liberal victory at the polls, he continued to wage fierce guerrilla warfare against the Marines and the Guardia Nacional.[5]

At the time of the first Liberal revolt, in May 1926, the Tommy-gun apparently was not in use by either the rebels or the first Marine landing parties sent to protect Americans and their property in the areas where fighting was going on. By the end of the year, however, the rebels had acquired a supply of Thompsons from sources in Mexico, and were using them to good advantage. In December 1926 the Liberal forces used the gun, for probably the first time, in a successful attack on Conservative forces at Pearl Lagoon, on the east coast of Nicaragua. General Delgadillo and about a thousand government troops had thrown up a star-shaped earthwork at a point on the lagoon where they intended to stop the Liberals from reaching the important port of Bluefields by boat. According to a Marine Corps account:

> On Christmas Eve, 1926, Escamilla, the Mexican then with the Liberals, opened firing with a battery of machine guns against the earthwork, from a thickly wooded hillock, about 400 yards distant. His fire raked the top of the earthworks and kept the Conservatives down behind them, while "General" George Hodgson, a Bluefields negro, in command of 12 negroes equipped with Thompsons, and each assisted by an ammunition carrier, advanced from a swampy thicket several hundred yards to the flank of the earthworks. The Conservatives were still held down by the machine gun fire, which was shifted to the other flank of the redoubt when Hodgson's Thompson gunners reached it on their side. The Thompson men lay on the parapet and thrusting the muzzles of their Thompsons over the top, poured in a murderous fire on the

surprised defenders massed within. Of the nearly 800 Conservatives actually present in the redoubt when the attack began, nearly 250 left their bodies on the field, while the rest jumped over the rear into a supposedly impenetrable swamp. Of these, several hundred lost their lives in the swamp, about 350 managed to get through it to safety.[6]

The news that the rebels had Thompson guns apparently had a demoralizing effect on the Conservative troops. A few weeks later a small guerrilla force armed with four Thompsons and eight .45-caliber Colt automatic pistols managed to capture a government outpost, partly by intimidating the defenders. According to the Marine report, ". . . the sound of the '12' Thompsons was enough; the Conservatives evacuated the position without delay."[7]

In January 1927, Fifth Regiment Marine reinforcements arrived bringing with them sixty-five of the Post Office Thompsons. The guns quickly proved invaluable in jungle combat, and by the end of the year another seventy-five had been acquired. In February 1928, the brigade commander in Nicaragua placed a rush order for an additional 200 Thompsons to raise the total to 340.[8]

The Marines in Nicaragua came to cherish the Tommygun and began urging its official adoption as Marine Corps standard issue. In a 1930 issue of the *Marine Corps Gazette* Captain Roger Peard, making such an appeal, described (though not too coherently) his experiences with the gun in the Nicaraguan jungle while pursuing Sandino's "bandits":

> The winding, wooded Nicaraguan trails, through very rough and mountainous country, where deep ravines, ridge trails, only wide enough for one man or animal to pass in single file, impenetrable forests, and impassable swamps, usually bordered the almost impassable trails, knee deep in mud, made the safe advance of our troops by the normal method of flank patrols for security, entirely impracticable or impossible. This left only one safe mode of advance practicable; to lay down a sweeping automatic fire on each possible enemy ambush location, at such range as to prematurely force the bandits to disclose their presence, and thus deprive them of the advantage of surprise, in their carefully prepared ambushes.

A short burst from a Thompson gun under such conditions, always drew a return fire from an ambushed enemy, and at such a range as to give our better marksmen all the advantage. Many ambushes were thus prematurely disclosed, as the bandits when thus fired upon always thought they had been discovered and immediately returned the fire, oftentimes when we did not think anyone was anywhere near, the protective fire having been simply laid down as routine, by members of the advance guard, on a dangerous location along the trail.[9]

The great popularity of the Tommygun in Nicaragua, and in China too, moved the Marine Corps to adopt the submachine gun as standard issue—one per rifle company squad of eight men. Much sooner than the Army, the Marines recognized the Thompson as an infantry weapon, and as a weapon in a class by itself. According to a Marine Corps manual published in the 1930s:

> The Thompson gun is one of the necessary infantry weapons, but it does not supplant any of the others. It has its own characteristics and its own place in battle. It is essentially a short range arm and particularly valuable at close-quarters when its ability to stop a man instantly and its ability to remain in action by use of the 50 round drum make it most effective. In fact its lack of penetrative power is converted into an advantage in street fighting or at close-quarters as it reduces the chance of injuring an innocent bystander or one of our own men.[10]

The Marines gave the Thompson its first real test as a military weapon, and also gave the gun a measure of respectability that it sorely needed toward the end of the "Roaring Twenties." In addition, Marine enthusiasm for the submachine gun awakened the Navy to its potential as a boarding and landing party weapon—something Marcellus Thompson had been trying to do for several years.

In 1920 the Navy had sent one of its ordnance officers, Lieutenant E. P. Donnelly, to witness the Army's tests of the Thompson at the Springfield Armory. In his official report Donnelly stated:

> The rate of fire claimed for this gun . . . is . . . 1800 shots per minute. They [Auto-Ordnance] contended that with the present gun they had fired 250,000 rounds without appreciable wear on the moving

parts and that they had fired 100,000 rounds without malfunction. It is believed that this arm would be suitable for landing forces and for street riot work and that it would be advantageous to the service to have the representative of the Auto-Ordnance Co. submit this gun for further test by a Navy Board.[11]

The Navy purchased several Thompsons for "experimental use" by the Marine Corps, but conducted no official tests, on the grounds that it was Navy policy to follow War Department recommendations with regard to small arms. In 1925, when Auto-Ordnance suggested the arming of Yangtze River patrol boats with Thompsons, the Bureau of Ordnance replied that, since the Army had thoroughly tested and rejected the submachine gun, the Navy had no interest in it.

In 1927, however, about the same time the Marine Corps procured its own submachine guns for use in Nicaragua, the Navy had a change of heart. That year it purchased a hundred Thompsons, allocating them to the river gun boats *Guam, Tutuila, Oahu, Luzon, Mindanao*, and *Panay*; to the aircraft carriers *Lexington* and *Saratoga*; and the cruisers *Rochester, Denver, Cleveland, Galveston*, and *Tulsa*.[12] Moreover, the Navy indicated it likely would adopt the submachine gun if slightly modified.

The Navy considered the Thompson's firing rate too high and its pistol foregrip too fancy. For the sake of accuracy and economy of ammunition, the Navy wanted the firing rate reduced from eight or nine hundred rounds per minute to about five hundred, and wanted the foregrip replaced by a horizontal forearm with carrying sling. The latter change posed no problem. Auto-Ordnance merely slimmed and lengthened the forearm used on the experimental Model 1923. To see what could be done about the firing rate, John Thompson, in March 1927, wrote a personal letter to Oscar Payne in Worcester, Massachusetts, asking his views on the subject. Payne, then designing weaving machinery for the Crompton & Knowles Corporation, answered that he probably could work something out, but it would have to be in his spare time. Thompson was agreeable to this, and Payne spent his evenings and weekends for the next three months devising a way to slow down the Tommygun's action.[13]

The solution Payne came up with was appealingly simple. By increasing the weight of the actuator (which effectively increased the inertia of the bolt), he managed to reduce the rate of fire to about six hundred rounds per minute, which was good enough to satisfy the Navy. Equally important, the new actuator required only minor alterations to the gun's internal mechanism: a recoil spring of smaller diameter, a new buffer and pilot, and (nonessentially) a slightly shorter firing-pin spring. All other parts remained the same.

The Navy officially adopted the Thompson, so modified, in 1928, which event Auto-Ordnance celebrated by marketing the gun commercially that same year as the "U.S. Navy Model of 1928." This was accomplished simply enough by taking Model 1921 Thompsons out of stock, substituting the new parts, and stamping them "U.S. Navy" with an "8" superimposed over the "1."

The 1928 "Navy" Thompson came with a Cutts compensator as standard equipment. Introduced optionally two years before, the compensator was a type of muzzle brake, about two and a half inches long, threaded onto the front of the barrel. By means of slots cut in its top half, it redirected a portion of the muzzle blast upwards, thereby reducing both recoil and the barrel's tendency to climb during full-automatic fire.

The Cutts compensator was the invention of Colonel Richard M. Cutts in collaboration with his son, Lieutenant Richard M. Cutts, Jr., both officers in the Marine Corps. The father and son team developed the compensator shortly after the First World War and submitted it to the Army about 1920, with a view to employing it on rifles, particularly the BAR. With high-powered rifle ammunition, however, the compensator created a disagreeable side and back blast, and for this reason the Army turned it down.[14]

Early in 1926, after reading newspaper accounts of the submachine gun's adoption by Chicago gangsters, Lieutenant Cutts had the idea of offering the compensator to Auto-Ordnance. (McErlane's mob, he may have noted, had been getting some bad groups with their Thompson.) Auto-Ordnance tested the compensator and found that it worked quite well with low-powered .45-caliber ammunition, increasing the gun's full-automatic accuracy as much as 140 per cent.[15] The company signed a royalty agreement with Cutts and introduced the compensator in the fall of 1926, promoting it as a recommended Thompson accessory, $25 extra. With compensators, the Model 1921 Thompson sold for $200 and the Model 1928 for $225, one Type XX (20-round) magazine included.

About the same time the Model 1928 was under development, but before it was marketed commercially, Auto-Ordnance introduced another modified version of the Thompson. This was the Model 1927 "Semi-Automatic Carbine." Like the 1928 model, the Thompson "Carbine" was merely a Model 1921 gun altered and re-marked. Its only new feature, or nonfeature, was that it fired only semiautomatically, by virtue of alterations to its rocker and rocker pivot. According to Auto-Ordnance literature, it was "not practicable" to convert the semiautomatic gun into a submachine gun due to "differences in construction and adjustment."[16] Except in a few easily replaced parts, however, any such differences seem to have existed only on paper.

The purpose of the Model 1927 Thompson is nowhere explained either in Auto-Ordnance handbooks or in the company's catalogs, which list its price as $175 without compensator. It was developed, according to George Goll, when Auto-Ordnance learned that some police officials were afraid to put submachine guns in the hands of inexperienced officers "who might not know when to let go of the trigger."[17] Only a few were ever assembled, and fewer still were sold.

About 1926, after four bad years, Tommygun sales gradually started picking up. Increasing "gangsterism" stimulated the police market, and the gun's adoption by the Coast Guard, Marine Corps, and Navy resulted at last in a few military orders. Political trouble in several parts of the world brought an increase in foreign sales, particularly to Mexico. But the most promising market of all during this period was Belgium, which had been showing interest in the gun off and on since 1921.

Unlike most other countries, Belgium had been quick to recognize the submachine gun's merits as a military weapon. During its second series of tests in that country, in 1923, the gun's performance earned praise from the army's highest ranking staff officers. Prince Leopold himself fired the Thompson, making good scores on the hundred-meter range, and later he entertained both General Thompson and George Goll at the king's palace in Brussels.

Belgian ordnance officials conducted a lengthy study of the Thompson and decided that it warranted adoption. They found fault with the gun's pistol-like shape, however, and with the .45-caliber cartridge, which was not Belgian standard. If Auto-Ordnance would remodel the Tommygun and chamber it for 9mm parabellum, ordnance officials assured, the Thompson would be adopted as standard infantry ordnance and purchased in large numbers.[18]

Lacking its own machine-shop facilities, Auto-Ordnance turned to England's Birmingham Small Arms, Limited, which in 1921 had expressed interest in manufacturing the Thompson for the European market. Nothing had come of the first talks between the companies, mostly for want of any good sales prospects. But when approached again in 1925, Birmingham Small Arms was still interested, and the two companies arranged to collaborate in developing a modified Thompson submachine gun that would meet Belgian military requirements and hopefully interest other European countries as well.[19]

B.S.A. arms designer George Norman was assigned to work on the Thompson project, and over a three or four year period developed an entirely new line of submachine guns that bore little outward resemblance to the American Thompson. The B.S.A. Thompsons retained the standard Thompson Blish lock, bolt-actuator system, finned barrel, and compensator, but their shape was more that of a carbine. The earliest version, designated the B.S.A. Thompson Model of 1926, reportedly had neither barrel fins nor compensator and measured about thirty-two inches in length.[20] Eventually this gun was refined into the B.S.A. Thompson Model of 1929, which measured about four inches longer and did employ a finned barrel and compensator.

Prototypes of the 1929 model were built in caliber .45 ACP (Automatic Colt Pistol, the standard Thompson round), 9mm parabellum,

9mm Bergmann Bayard, 9mm Mauser, 7.63mm Mauser, and 7.65mm Luger. George Goll, who had acted as liaison between Auto-Ordnance and B.S.A., presented the 9mm parabellum model of the submachine gun to Belgian ordnance officials that same year. After a series of tests the Belgian government informed Auto-Ordnance that the new submachine gun met all military requirements for adoption, and that it would like to negotiate a licensing agreement for the manufacture of ten thousand B.S.A. Thompsons by the country's own national arms factory, Fabrique Nationale d'Armes de Guerre.[21]

Meanwhile, the long-struggling Thompson Autorifle was also showing new signs of life.[22] The rifle's early models, the A, B, and C, using one form or another of the sliding wedge lock, had been shelved in 1917. The sliding wedge had proved successful in the submachine gun, but too light and unreliable for use with high-powered rifle cartridges.

In 1919 Oscar Payne started work on two new versions of the Autorifle. One, the Model D, still employed the wedge principle, but the wedge was essentially rod shaped, and in line with the bolt, with contiguous surfaces meeting at an angle so as to cause the locking member to rotate (instead of slide) into its unlocked position. This system soon was abandoned, however, in favor of another type of rotating lock, inspired by Canada's Ross bolt-action rifle. The Ross rifle's hand-operated bolt locked on interrupted threads, as did the large naval guns that originally inspired John Blish. Experimenting with the Ross, Payne and Eickhoff discovered that by removing some of the locking cams, or threads, the bolt could be made to rotate, unlock, and open under normal chamber pressure. This eliminated the need for a separate locking piece of any kind, and since the inclined thread itself constituted a "wedge," it remained a Blish lock in principle.

Designated the Model P, in honor of Payne, the new Autorifle underwent tests at the Springfield Armory in 1920. It passed the 5,000-round test with no breakage or repairs, setting a record for semiautomatic rifles, but in other respects it failed to measure up to the Berthier machine rifle or to the experimental rifle designed by

J. C. Garand. In 1921 Auto-Ordnance submitted an improved version of the rifle, the Model PC (for Payne and Colt's, whose engineers also had worked on it), as well as one of the earlier Model C sliding-wedge rifles. The Model PC also failed to meet Army requirements, and the Model C ejected its shells with such force as to constitute a hazard. Julian Hatcher, the Army's well-known firearms expert, tested the rifle and found some of the spent cartridges stuck in a wooden door twenty feet away.

Auto-Ordnance improved the Model PC, and in 1923 the Cavalry Board found the Autorifle (now designated the Model of 1923) to be simple, accurate, rugged, dependable, and correctible in its remaining faults. In 1925 the Ordnance Department ordered twenty Thompsons for field trials by Army regulars. In December of that year, the Infantry Board at Fort Benning subjected the Autorifle to extensive tests in competition with the semiautomatic developed by Garand at Springfield. At the conclusion of the tests the following spring, the Ordnance Department gave out no official report—possibly because some members of the department still were strongly opposed on principle to the adoption of any semiautomatic rifle. However, the writer of a military column in a Springfield newspaper disclosed that

> . . . the Garand gun lost its chamber pressures on occasional shots, and that its accuracy was not entirely satisfactory; that the Thompson gun more than held its own in accuracy, and, as a matter of fact, proved superior to the Springfield [M1903] in that respect; and that the Thompson mechanism, as displayed by the latest models before the board, established an actual physical saving in weight of over one pound. . . .
>
> One of the subjects that was considered during the firings at Fort Benning relates to automatic oiling, and it is understood that it was admitted by at least some members of the board that oiling is the only solution of the problem of obtaining uniform breech openings with corresponding accuracy. Indeed, it is reported that the Thompson type oiling pads [installed in the magazine] were placed on the Garand gun.[23]

About the same time, Auto-Ordnance commissioned Birmingham Small Arms to build an Autorifle in caliber .303 for entry in the

British government rifle tests of 1928. The Thompson beat the entries from several countries, including England, and won the British War Office prize of $15,000 as the best semiautomatic rifle submitted. This qualified the Autorifle for general trial among British troops, with a view to its possible adoption as the standard British army service weapon.[24]

With domestic sales picking up, with the submachine gun's adoption by the Coast Guard, Navy, and Marine Corps, with the prospect of a large Belgian contract, and with the Autorifle's successes in the United States and its excellent reception in England, the year 1929 looked to be the long-awaited turning point in Auto-Ordnance fortunes. And it was, but not in the direction that John and Marcellus Thompson were hoping.

Before the Belgian government appropriated the funds to put the submachine gun into production, the Depression struck, and national economies collapsed. Military spending stopped. The B.S.A. Thompson never progressed any further than the prototypes. Nor did the Autorifle, presumably for the same reason. In the United States Tommygun sales, which had been climbing since 1926, started falling off again.

Worse even than the Depression in some ways was another disaster that befell the company and the Thompsons in 1929. Thomas Fortune Ryan had died at his palatial Fifth Avenue home in Manhattan on November 23, 1928, and his controlling interest in the Auto-Ordnance Corporation had passed to his estate. The estate's executor, Elihu Root, the famous statesman and internationalist, disliked machine guns on principle, and the Ryan family had never much approved of their patriarch's dabbling in munitions. Anxious to liquidate this undesirable holding, the Ryan estate, in 1929, foreclosed the chattel mortgage notes securing Ryan's cash advances to Auto-Ordnance since 1916. This reduced the underlying property to ownership of the estate, which made all outstanding shares in the corporation valueless to the minority stockholders. Next the estate, in 1930, forced Marcellus Thompson out of office as vice president and filled the board with members and associates of the Ryan family.[25]

Elihu Root and the trust department of the Guaranty Trust Company installed as president Walter B. Ryan, Jr., a nephew of Thomas Fortune Ryan, who undertook to salvage what he could of the Ryan investment and dispose of the company. This investment was considerable. Before his death Thomas Fortune Ryan had advanced Auto-Ordnance a total of $1,312,954 for developmental, patenting, and manufacturing expenses. During the period from May 1921 to December 1930 gross sales totaled approximately $1,200,000, of which only $50,000 had been used to reduce the company's obligations to the North Virginia Corporation, Ryan's holding company.[26]

Walter Ryan's first efforts were to sell Auto-Ordnance outright at a bargain price rumored on Wall Street to be $650,000. This amounted to offering the entire inventory of some eight thousand guns at about 60 per cent off list price and throwing in free the company and its assets. When the Thompsons, as minority stockholders, threatened court action to stop such a move, the Ryans relented and settled back to preside leisurely over the company's gun-by-gun liquidation.[27]

Auto-Ordnance developmental expenses through December 1921 had totaled $355,900. After 1921, developmental expenses amounted to $21,408 (1922), $30,902 (1923), $26,619 (1924), $22,769 (1925), $15,782 (1926), $1,576 (1927), $4,505 (1928), and $27,043 (1929). In 1930, the year the Ryan estate took control of the company, development costs fell off to about $600, and the next year to zero.[28] Walter Ryan simply ordered all developmental work discontinued and vetoed Marcellus Thompson's plans for any further commercial expansion. In addition, he instituted a new and highly restrictive sales policy aimed at getting the now-notorious Tommygun out of illegal circulation. On September 5, 1930, Auto-Ordnance issued the following circular letter to the company's wholesale and retail outlets:

> On rare occasions our gun has gotten into the hands of enemies of society. This has happened notwithstanding our earnest efforts of jobbers and salesmen who have handled guns for us. The fact that a New York newspaper stated that the price offered by bandits in Chicago for a sub-machine gun is between $500 and $1000 is clear evidence that the

gun is not readily available to crooks. We feel that in fairness to the police and notwithstanding the non-existence of laws restricting the sale of sub-machine guns, the gun is so deadly and effective that we ought to leave no stone unturned to stop all leaks. We have therefore established the following policy for the better check-up of the ultimate destination of all guns sold by us.

FIRST: We sell only for use by officers of the law—such as Army, Navy, Marine Corps, police, sheriffs, penal institutions, militia, troopers, and corporations or institutions clearly entitled to have the gun for protection such as Armoured Car Corporations, Federal Reserve and other duly established banks.

SECOND: We sell only on receipt of written order from the ultimate purchaser which order must set forth that the gun is purchased for the use of the department. In some instances in case of immediate necessity, we will ship upon exchange of telegrams or telephone call with the officer requiring the gun.

THIRD: We have appointed agents in some states in which all sales must come through our representative. This is the case in Ohio, Indiana, and Michigan where we are represented by E. E. Richardson of Maumee, Ohio. All orders given to our representatives are, however, subject to approval by an executive officer of the company.

We trust that none of our customers or representatives will chafe at these restrictions and that the relationship which we have built up will continue unimpaired by these rules which are meant for our mutual protections and to keep our sub-machine gun always *on the side of law and order.*[29]

A few weeks later, Ryan virtually closed down the sales department. In a circular letter issued October 22, 1930, the same day the Ryan estate assumed full control of Auto-Ordnance, Walter Ryan informed all "dealers, jobbers, and salesmen" that:

The management of the Company has decided to discontinue, for the time being at least, all sales of Thompson Sub-Machine Guns for other than military purposes. Consequently, no orders will hereafter be accepted from the police, sheriffs, prisons, banks, or other non-military purchasers.[30]

When the minority stockholders obstructed the estate's plans to dispose of the company, Walter Ryan resumed selling guns to law enforcement agencies. To cut out the middleman, to whom cash often meant more than a customer's character, Auto-Ordnance opened a modest direct selling campaign aimed only at federal, state, and municipal police agencies. Orders from banks and protective agencies were accepted only after each had been specifically approved by the office of the United States Attorney General.

In formulating the new Auto-Ordnance distribution and sales policies, Ryan, apparently on his own initiative, had gone to the Attorney General's office for advice. Later he vigorously supported the proposed federal firearms law when every other firearms manufacturer and every sportsman's organization was fighting it tooth and nail. During hearings on the bill, Ryan testified: "We have studied the bill fairly carefully and we believe that the provisions of it will materially aid in the disarming of the criminal. The policies of the company itself have been exactly those as embodied in the pending bill for a number of years."[31]

This was a radical departure from the Auto-Ordnance attitude toward firearms legislation under Marcellus Thompson. In 1926, when asked by a magazine writer if there were no way to keep submachine guns out of criminal hands, Marcellus voiced the arms maker's traditional plea of helplessness: "Well, I don't know any way of stopping it. What can I do? Here's a new invention; we must have progress in arms for military purposes and for public safety just as we have progress in every other mechanical line."[32] In 1928 Auto-Ordnance opposed a bill to outlaw machine guns in New York State, on the curious grounds that it was "motivated by commercial favoritism for the manufacturers of such arms as sawed-off shotguns."[33]

Ryan's program of direct, supervised selling to police agencies failed completely, and by 1932 Tommygun sales had reached a new low of less than ten orders a month. At this rate it would have taken sixty or seventy years to exhaust the Auto-Ordnance inventory. To rush things up a bit, Ryan turned sales over to an agency, Federal Laboratories, Incorporated, of Pittsburgh, which on April 4, 1932, became the Thompson's exclusive supplier in the United States and its major distributor to foreign countries.[34]

A "protection engineering" firm, Federal Laboratories was (and still is) the country's principal supplier of tear gas, gas guns, and riot and police equipment in general. As such, it was in direct contact with virtually every law-enforcement and private protective agency in the United States. Through Federal, Auto-Ordnance could sell guns to police and other legitimate purchasers without going through either wholesale or retail dealers, or otherwise putting the Thompson on the commercial market.

This new arrangement did much to stop the Tommygun from reaching bona fide gangsters, but it soon dragged Auto-Ordnance and the Thompson into a new series of scandals. Federal's salesmen proved eager to a fault, and Senate investigators soon were accusing the company (sensationally, if not always accurately) of gun smuggling, meddling in Latin American politics, fomenting labor violence, evading the National Firearms Act, and generally playing the *agent provocateur* to drum up tear gas and Tommygun business.

During the years of social readjustment following the World War, American businessmen and industrialists felt increasingly threatened by Bolshevism, anarchism, and most of all, unionism. Organized labor had made measurable headway during the war years, and the efforts of unions to preserve their gains seemed doubly menacing in the context of other events: the Russian Revolution, the Wall Street bomb, the Sacco-Vanzetti trial, the Palmer "Red raids," the virtual massacre of some twenty-five "scabs" during a coal strike at Herrin, Illinois. Several violent strikes only confirmed capitalist fears, creating a demand for strike-breaking agencies and riot equipment, including submachine guns. After a few years of relative peace, the Depression spawned a new era of labor-management conflict. Many large industrial firms refused to concede the "right to organize and bargain collectively," which became law with the National Industrial Recovery Act of 1933, and fought union "agitators" with tactics that amounted to simple terrorism. The NIRA was ruled unconstitutional in 1935, but its important labor provisions were preserved, to the dismay of management, in the new National Labor Relations Act of that same year.

In the spring of 1936, after three years of violent strikes and their violent suppression, a Senate subcommittee headed by Robert M.

La Follette, Jr., opened an investigation into violations of the rights of labor and free speech. The subcommittee's disclosures were both grim and sensational: large industrial corporations spending hundreds of thousands of dollars on tear gas, vomiting gas, and submachine guns, and hiring spies to infiltrate unions and gunmen to intimidate and sometimes kill union organizers; "protective" agencies supplying weapons and recruiting small armies of professional thugs to break up strikes and union meetings.

Federal Laboratories figured prominently in the investigation, partly because some of its young salesmen possessed a sense of humor that failed to amuse the stodgy Senators. One salesman jokingly complained in a letter to the home office, "I think someone should get out a restraining order on the President of the United States to prevent him from stopping all these strikes. It seems to me that his actions are absolutely in restraint of trade—that is, so far as we are concerned." Another wrote of "nice juicy strikes," complaining that the "darn things don't happen often enough to suit me. . . . I honestly believe I can join the Ancient Order of the Ghouls pretty soon."[35]

The investigation disclosed that Federal Laboratories was in cahoots with the Railway Audit & Inspection Company, a large strike-breaking agency, and had armed R.A.I. "deputies" with submachine guns through the sheriffs and police chiefs of company towns. The police officials would order the guns, pay for them with company funds, and then deputize company-hired R.A.I. guards so they could legally carry them.

On September 6, 1934, Federal Laboratories telegraphed a rush order for four Model 1928 Thompsons to Auto-Ordnance on behalf of West Point, Georgia, Police Chief W. E. Boyd. Walter Ryan filled the order, but notified the Treasury Department that payment for the guns had been made by the West Point Manufacturing Company. A Treasury agent investigated and discovered that the guns had been turned over to a West Point subsidiary, the Lanett Mills of Lanett, Alabama, where a strike had been called. The Lanett Mills superintendent had issued the guns, along with three Thompsons borrowed from the DeBardeleben Coal Corporation of Birmingham, to company guards deputized by the local sheriff.[36]

Senate investigators heard testimony from Ester Lee Groover of

Lanett, who owned a small dry-cleaning business and openly sympathized with the union. Groover described his being lured to a rented room to pick up some cleaning, and there being beaten up by company guards armed with two Tommyguns. Groover related two other incidents he had witnessed. On one occasion the guards had machine-gunned the tires of a doctor's car at a roadblock, and another time used their Tommyguns to hold back an angry crowd of workers while they beat up a fifty-four-year-old man, J. R. Hamby, for "agitating" and criticizing the guards.[37]

Although he had no sympathy for Reds, labor agitators, and other un-American types, Walter Ryan did try to keep Tommyguns from reaching strike breakers. In October 1933 a Federal Laboratories official wrote the Rex Coal Company in Chicago that he had been

> . . . working hard on the question of affecting delivery of the Thompson submachine gun you recently ordered through us. Auto-Ordnance has a fixed rule that they will not ship guns to corporations where these guns are specifically to be used in the breaking up of strikes. In the past so much furor has been caused when the guns have been so used, and the federal authorities have made so much noise about it that they dare not and will not ship guns for that purpose.
>
> If you want this gun and are willing to have the Sheriff's office or the Police Department in your locality officially order it and they will give us an affidavit to the effect that the gun will not be transferred without written permission from the manufacturer, then the gun can be shipped, and if they want to lend you the gun there will be no way, of course, that exception could be taken to such a transaction.[38]

In fact, both Auto-Ordnance and the Treasury Department took grave exception to any "lending" of machine guns, and quickly put a stop to such practices once they became known. But by the middle Thirties hundreds of machine guns had already been sold to industrial firms, and a thousand or more to collaborating police officials, to the further discredit of General Thompson's "trench broom."

Federal Laboratories suffered considerably at the hands of the La Follette committee for its shady role in domestic labor conflicts. For its questionable export activities the company had got itself (and its

Tommygun) in equally hot water with the Nye committee, investigating the nefarious activities of the munitions industry.

About 1933 came a veritable flood of books, articles, and orations attacking the world's munitions makers as "merchants of death" who fostered international conflict and precipitated wars in order to reap enormous bloodstained profits. The venerable old Colt firearms company, it was charged, had begun tooling up for the World War as early as 1908 and had worked vigorously over the next six years to bring it off. DuPont was accused of having profiteered to the extent of $250,000,000, after expenses and taxes, on powder sales alone. With the eruption of a bloody war in the South American Chaco, Latin America generally in turmoil, and rumors of German rearmament, the United States Senate appointed a special committee headed by North Dakota isolationist Gerald P. Nye to inquire into the presumed machinations of the munitions makers.

The Nye committee's disclosures were predictably thrilling: governmental bribing, mislabeled cargoes, powerful motor launches running crates of guns into fog-shrouded coves by moonlight. Sensational (if largely circumstantial) evidence indicated that munitions manufacturers were arming one country, then informing neighboring countries to induce them to arm as well. As usual, Federal Laboratories found itself in the middle of things. Reported the *New York Times* in 1934: "The Committee also began today its inquiry into the domestic and foreign activities of American manufacturers and dealers in military gases and machine guns, especially Thompson guns."[39]

In the files of Federal Laboratories the investigators turned up a good deal of interesting correspondence that seemed to exhibit the arms maker's traditionally cavalier attitude toward war. In the summer of 1932 Federal's sales representative in Buenos Aires, with the usual flair for tactless wording, reported on his diligent efforts in Chile, Peru, Bolivia, Paraguay, and Uruguay:

> While we cannot expect, nor do we expect, more on account of the deplorable financial conditions of these countries to get immediate results, still the fact remains that there is a live interest, and we are

keeping after the heads of these governments. . . . You are no doubt acquainted with the bellicose conditions which exist between Bolivia and Paraguay. . . . For the last year or more there has been a guerilla warfare going on in the Paraguayan Chaco. . . . Last week the Bolivians and the Paraguayans were in trouble again, and the general opinion in these countries was that a declaration of war was imminent.

Taking advantage of this state of affairs, the writer has insistently pushed both the Bolivian and Paraguayan Governments.[40]

Enthused by this report, Federal rushed one of its men to Washington to visit both the Paraguayan and Bolivian legations. The results, however, were disappointing. "Unfortunately, for us," the agent reported, ". . . it looks as if the trouble they are having is going to be settled amicably."[41]

Happily, the agent was proved wrong by the outbreak, a few months later, of the exceptionally bloody three-year-long Chaco War, during which Federal Laboratories did a thriving business with both countries. "We are not discriminating," Federal's president, John W. Young, told the Senate investigators.[42]

Federal's Cuban activities were even more questionable. In 1932 Federal was urging the Machado government to "speed the placing" of orders for gas and Tommyguns to deal with the impending revolution, and at the same time negotiating with the rebels. Machado fell in August 1933, and the new regime rewarded Federal Laboratories with letters of praise, a contract to set up and equip a new national police force, and orders for gas and Thompsons exceeding $400,000 in the first year.[43]

Political chaos gripped Cuba throughout the early Thirties, and the Tommygun became a popular weapon of assassins and terrorists, as well as the military. During one period of rioting in August 1933, carloads of gunmen in civilian clothes roamed Havana, indiscriminately firing into crowds with machine guns and shotguns. A witness described one incident to a *New York Times* reporter:

I was standing at the corner of Castillo and Monte Streets when people began to rush into the street shouting [at a false report of Machado's fall]. A Cuba Company omnibus filled with cheering chauffeurs came

> down the street. At that moment a red automobile . . . came around the corner and sprayed the omnibus with machine-gun fire, wounding all of the occupants. Some of them must have been killed. Two youths who were running along the opposite sidewalk were shot and killed instantly. The machine disappeared down the street, firing from both sides.[44]

The following October another Tommygun massacre took place in Havana. Rebel soldiers guarding a large group of deposed army officers outside the National Hotel opened up on their prisoners with submachine guns after a sniper fired a shot from a hotel window. At least ten officers and eight civilians were killed, and forty-six others were wounded. In June 1934 a carload of gunmen attacked a political parade with machine guns, killing fourteen and wounding sixty.

During the late Twenties and the Thirties the Thompson managed to reach political groups, whether in or out of power, all over the world. Tommyguns sold legally to the Mexican government during the Twenties were answered by Tommyguns issued to American soldiers of fortune fighting for the rebels. In 1929, American adventurers flying planes for General Escobar's revolutionaries used Thompsons in dogfights with government planes during the battle of Jiménez. In 1935 Mexican military aircraft were reported damaged by "Tommygun" fire from the ground as they reconnoitered a rebel camp in Sonora.[45]

Around the end of the Twenties, the Thompson also made its way back into Ireland, where die-hard remnants of the old I.R.A. had resumed their sporadic efforts to unite the country by means of the sword. Eamon de Valera had split the Republican movement in 1922 by leading the radicals into a civil war against the "traitors"—those Republicans who had assented to the division of Ireland into the Free State and Ulster, or Northern Ireland. Weary of bloodshed, de Valera turned from bullets to ballots in 1927, only to find himself now labeled a sellout by the radical faction of his own followers, who vowed to continue the fight on their own. In the early Thirties the Kerry football team managed to smuggle in a number of Thompsons in their luggage, and about 1934 some three to four hundred were slipped into Cork via Galway. Small lots of Thompsons reached the rebels throughout the Thirties, figuring in occasional border raids against Ulster and attacks on military armories that have continued (if on a rather amateurish level) into the 1960s. Despite the Thompsons that the authorities have captured or confiscated, which probably number in the hundreds since 1921, the gun seems to have become a permanent part of Ireland's underground arsenal. Eight captured Tommyguns were handed over by the British to the Free State army following the 1921 truce; 15 were surrendered by the public in response to a government appeal for arms in 1940; and 106 were found in a secret County Mayo arms dump in 1942. In 1954 and 1955 I.R.A. irregulars left Thompsons behind in unsuccessful attacks on an armory and a police station. In December 1956 the commanding officer at Armagh, who happened to be in town, knew that the army barracks were under I.R.A. attack by the familiar sound of an old-fashioned Thompson firing off in the distance.[46]

No large number of Tommyguns reached Germany at any one time, but in June 1932, just before Hitler came to power, the DuPont sales representative in Paris passed the following intelligence along to his employers in Wilmington: "I heard on my last visit to Holland that the German political associations, like the Nazi and others, are nearly all armed with American revolvers and Thompson machine guns."[47] Six months later the same salesman elaborated in another letter:

> There is a certain amount of contraband among the river shippers, mainly in arms from America. Arms of all kinds coming from America are transshipped in the Scheldt to river barges before the vessels arrive in Antwerp. Then they can be carried through Holland without police inspection or interference. The Hitlerists and Communists are presumed to get arms in this manner.
>
> The principal arms coming from America are Thompson submachine guns and revolvers. The number is great.[48]

"Transshipment" was the most common means by which arms exporters evaded embargoes and other trade restrictions. Guns could be sold legally to a dealer in some neutral foreign country and there resold, or merely forwarded, legally or otherwise, to their real purchasers.

Walter Ryan seems to have worked closely and conscientiously with the State Department in regard to foreign sales, and to have kept a close eye on his export dealers. At the same time, however, he put constant pressure on Federal and its affiliated exporters to sell guns. So they sold guns—above board when possible, under the table when they thought they could get away with it. In May 1932 Federal Laboratories export agent Frank Jonas wrote a colleague in Brazil that he had gotten wind of a Brazilian order for a million 7mm rifle cartridges, which indicated that a revolution might be brewing. He suggested that his contact try to get in touch with "the right parties, as you might be able to interest them in the Thompson gun. The Thompson is known in Rio Grande [do Sul] and has been used with success at various times."[49] A year later, in August 1933, Jonas received a rather candid order from another Brazilian agent on behalf of the local rebels. "The Revolutionary Government here is organizing a secret society and want machine guns and ammunition. The necessary contraband here can be arranged so long as they come in invoiced as machine parts. Later on they will need machine guns and ammunition officially. In short, the first lot will be secret; the second, official."[50]

Jonas sent his regrets, on the grounds that the deal was "too risky" and that Auto-Ordnance would not allow him to export any guns without permission from the State Department and from the embassy of the country to which the guns were going. He added, however, that he happened to know a fellow in New York who had a good stock of used Lewis guns and Browning automatic rifles.[51]

Walter Ryan's strict sales policies and the new state and federal laws did much to take the Tommygun out of circulation and out of action. No sensible gangster wanted such a hot item lying around the house when the police came calling, particularly since an ordinary pistol or shotgun could do essentially the same job. After 1934 the Tommygun remained in use only by a few status-conscious amateurs, such as the Al Brady gang, who presumably considered it a requisite to acquiring any measure of bank-robbing fame. According to one unverified account, the Brady gang even troubled themselves to steal an old World War I water-cooled machine gun (minus bolt mechanism) from an American Legion monument.[52]

In December 1934 the *New York Times* reported happily that not a single machine gun or sawed-off shotgun had been sold, except to law enforcement agencies, since the passage of the National Firearms Act, and that 15,791 such weapons had been registered with the Treasury Department. Two years later, in 1936, the *Times* reported that only one such weapon had been sold during the previous fiscal year. The paper this time, however, considered such a lonely statistic an indication that people were simply evading the law, since some twenty-five manufacturers and dealers were paying the stiff annual license fee for the privilege of handling such weapons.[53]

In addition to hampering criminals, Walter Ryan's sales restrictions made things hard for the Thompson's export agents at a time when revolutionary activity in the Caribbean and Latin America was creating a bull market in both legitimate and contraband arms. In May 1934 the Tommygun's foreign market shrank even further following a ban on munitions exports to Bolivia and Paraguay, then thoroughly engaged in the Chaco War. The result of these various restrictions was to leave the international arms market wide open for submachine guns of foreign manufacture.[54]

During the late Twenties and early Thirties European armament firms brought out a variety of "machine pistols," as they were called, mostly patterned on the original German Bergmann MP 18. The famous German gun designer Hugo Schmeisser, designer of the Bergmann, later refined it into the Schmeisser MP 28II for the C. G. Haenel Waffenfabrik firm in Suhl, Germany. In order to evade the terms of the Treaty of Versailles, which prohibited German manufacture of such weapons, Haenel licensed it to Anciens Etablissement Pieper in Herstal, Belgium, which introduced it commercially about 1928. Other German companies and designers likewise evaded the treaty by licensing manufacture of their submachine-type weapons to foreign firms, or by acquiring control of arms factories in foreign countries.

During this same period of general rearmament the Bergmann company marketed its own improved version of the MP 18, known as the SIG, after the Swiss Industrial Company (Gesellschaft) at Neuhausen am Rheinfalls, Switzerland, where it was manufactured under license. In 1929 the German firm Rheinmetall purchased the Waffenfabrik Solothurn, A.G., in Solothurn, Switzerland, for the purpose of manufacturing and marketing weapons of German design, including a line of submachine guns. Some Rheinmetall designs were also subcontracted to the Steyr-Daimler-Puch firm in Steyr, Austria. The machine pistols of Rheinmetall design were designated either Steyr or Solothurn, depending on the factory, and were marketed through a joint company, Steyr-Solothurn Waffen of Zurich, during the 1930s.

About 1930 the German firm of Erfurter Maschinenfabrik introduced the Vollmer-designed ERMA, and a year later Finland marketed its Suomi submachine gun, which was also sold to Denmark, Sweden, and Norway. About the same time, Spain developed the "Star," trade name of the Bonifacio Echeverria company in Eibar. In 1934 Belgium, which had flirted with the Thompson throughout the Twenties, adopted the simpler and cheaper Schmeisser MP 28II, designating it the Model 34 Mitraillette (the French and Belgian term for small-caliber full-automatic gun).

This rash of new "machine pistols" represented a radical change in European military thinking since the World War. Except in France and Great Britain, where faith remained strong in the Maginot Line and the English Channel, military planners were rejecting old-time fortification in favor of fire power and mobility. Except for France and Great Britain, virtually every country recognized and adopted the submachine gun as a new, separate class of military small arm.

The submachine guns manufactured in Europe bore little resemblance to the Thompson. Most were straight blow-back and patterned along Bergmann lines, with one-piece rifle-type stock and perforated cooling jacket around a short barrel. Compared to the Thompson they were crude in design, construction, and finish. But they were also comparatively simple and cheap to manufacture, and they were readily available on the international arms market.

The appearance of cheap, effective foreign submachine guns, and the possibility that they might reach the United States, alarmed Federal Laboratories. As the exclusive distributor of Tommyguns in the U.S., Federal had found the Thompson to be an excellent sales leader in pushing its entire line of police equipment. Any police department that wanted a submachine gun had to go to Federal, which meant that Federal also would get the department's order for tear-gas guns, gas grenades, and so on. This situation had prompted Federal's chief competitor, the Lake Erie Chemical Company of Cleveland, to try to market a rival submachine gun designed by George Hyde of Brooklyn. Unable to obtain the Hyde gun, which was still in the developmental stage, a Lake Erie salesman complained to his home office in 1934:

> I have lost sales for 37 machine guns alone in the past four months. . . . This sheriff had no money to even buy $100.00 worth of gas, although I have contacted him for over a year. Federal sells him machine guns and then gets an order for gas as an auxiliary to the machine guns. I have orders on my desk right now for sheriffs' and chiefs' friends who have held up their purchases for 30 days in an effort to give me the business, but gradually have had to give Federal the business on account of pressure of bankers and others that they get machine guns for protection. Every village here has gone gun crazy, and the only way gas can be sold now in the future will be with machine guns.[55]

The Hyde submachine gun outwardly resembled the Thompson, having a finned barrel, twin pistol grips, and an almost identical profile. It was even designed to take Thompson magazines. In 1933 or 1934 several of the Hyde prototypes were sold to the police chief of Panama City, Panama, who thought he was buying Thompsons because of the name on the drums. The Federal salesman who stumbled onto the gun in Panama wrote back describing it as "an imitation Thompson, so far as the barrel is concerned, and the breech looks like a bicycle pump."[56]

The Hyde gun never got into production, but by this time Federal also had started worrying that some American firm might begin importing or even manufacturing the German Schmeisser or some other European gun. This possibility had prompted Federal Laboratories to join Auto-Ordnance in supporting proposed federal laws that would regulate the importation and sale of submachine-type weapons. In the summer of 1933 John Young, president of Federal, wrote to Walter Ryan:

> We seem to have some new competition facing us in the Schmeisser machine pistol, a Belgian [actually German] development, which happens to be an improvement on the Bergmann submachine gun. . . . I have examined one of these guns which Lt. Cutts has. It appears to be smaller and simpler than the Thompson. It gives me some concern if it should reach the U.S. market. . . .
>
> Inasmuch as you were away, I took the responsibility upon myself to go before the Attorney General with the suggestion that an embargo be placed by the President on the importation of all submachine guns in the U.S., pointing out that this action would be necessary if they hoped to make any progress in their drive against crime. This suggestion was well received and is being passed on to President Roosevelt this week. . . . Assuming the President will take the requested action, we have only to consider then the matter of American manufacture under patent rights. Harrington & Richardson or Sedgley[57] may take some such rights and develop competition for us on a price basis that would prove embarrassing.[58]

European submachine guns did reach the United States during the Thirties, but not in numbers great enough to influence Thompson sales or the national crime situation. Elsewhere, however, they quickly squeezed out the Tommygun. Owing partly to trade restrictions, but more to lack of any serious sales efforts by Auto-Ordnance, the fighting in South America, China,[59] Spain, and Africa was done with "machine pistols" while several thousand Thompsons gathered dust in a Hartford warehouse.

8

The Road to War and Profits

I have given my valedictory to arms, as I want to pay more attention now to saving human life than destroying it.

—*John T. Thompson, 1939*

The death of Thomas Fortune Ryan and the consequent change in Auto-Ordnance management, together with the Depression, wrecked the Thompsons' hopes that the submachine gun's hour finally had come. To compound their disappointments, these setbacks occurred at a time when John and Marcellus Thompson were both undergoing trials in their personal lives.[1]

Marcellus Thompson's marriage to Dorothy Harvey had been less than a happy one. The handsome young society soldier had turned into a stolid, not-too-successful businessman, while the glamorous and somewhat pampered Washington debutante remained just that. After seven and a half years, the fragile relationship began breaking up over an interfamily incident—Thompson's indictment in the Irish gun-smuggling plot. Whether or not Marcellus was personally involved in the conspiracy, he must have known or guessed what was going on, and George Harvey blamed his son-in-law for the embarrassment caused him as ambassador to Great Britain. Had the plot succeeded, it would have jeopardized British-American relations, not to mention Harvey's diplomatic career. After having helped the Thompsons secure financial backing for the gun, he felt betrayed. His warm feelings for the Thompson family cooled considerably, and his daughter, Mrs. Marcellus Thompson, sided with him and against Marcellus.

In 1923 Dorothy returned to newspaper society pages by visiting her parents in London and attending the wedding of the Duke of York by special invitation of King George. When Ambassador and Mrs. Harvey returned to the United States, Dorothy remained in London for a time as the guest of the American chargé d'affaires. She made society news again when the Prince of Wales selected her as his dancing partner at a gala ball given by the French ambassador. Some time after her return home she and Marcellus separated. They were divorced on February 9, 1929, and a week later Dorothy married Augustus Smith Cobb, a vice president of New York's Bankers Trust Company who had been divorced in Switzerland the previous month. About 1932 she was taken ill with tuberculosis. Her condition worsened steadily, and in the early morning of April 4, 1937, in her eighteenth-floor suite at the St. Regis Hotel in New York, she shot herself in the head with a small automatic pistol.

John Thompson was saddened at the breakup of his son's marriage, and at the split between the Thompson and Harvey families that had led to it. The gunrunning scandal not only had caused him great embarrassment; it cost him the friendship of George Harvey and the frequent company of his little granddaughter, Dorothy Marcella, whose pictures, letters, and clippings received prominent display in his personal scrapbook.

In September 1929 Marcellus married Evelyn Allensworth, the daughter of a Hopkinsville, Kentucky, lawyer. They soon had two children—Evelyn Southgate and Juliet Ferguson Thompson, born in 1930 and 1931. The arrival of a new grandchild in the summer of 1930 delighted John Thompson, but much of his pleasure was offset by the death of his wife only a few weeks later.

Juliet Estelle Thompson died at the age of seventy-two in the white frame house on New Canaan's Bank Street where she and her husband had lived since 1919. She had been Thompson's close and constant companion in forty-eight years of marriage, living wherever he was stationed, often going with him on his travels in the United States and Europe. She was well known and liked in Army social circles and had worked energetically with relief agencies during the World War and the Spanish-American War. By nature she was a charming and extremely charitable person, "continually doing good deeds in a very quiet way."[2] Her death was mourned not only by John Thompson and Marcellus, but almost equally by Eickhoff, Goll, and Payne, to whom she had been a "second mother" when they were young employees at Auto-Ordnance. John Thompson expressed his own grief in a short letter to Oscar Payne:

> Marcellus and I deeply appreciate the beautiful flowers & your understanding sympathy—for you too have suffered a great loss. I miss my sweetheart & old pal. It is so terribly lonesome without her. We had over 48 years of happiness together. As a girl she had a great horror of death in all its aspects. She overcame that fear, for [while I was] holding her dear hand in her last moments of consciousness she said, "I am not afraid to die. I am ready." Then fixed her eyes on her picture of Christ on her bedside wall & tried to repeat her "creed," but faded away.
>
> I enclose a clipping from the local paper & also a copy of her favorite creed taken from her little "Unity" book—
>
> I shall carry on as bravely as I can. God bless you all.
>
> Devotedly your old friend,
> John Taliaferro Thompson[3]

The same year his wife died John Thompson retired from the Officer Reserve Corps, his wartime rank of brigadier general converted from Reserve to Regular Army by a special act of Congress. He moved from New Canaan to Great Neck, Long Island, to stay with Marcellus and Evelyn Thompson at 17 North Drive. There he lived quietly, busying himself with household affairs and inventing things on paper. Occasionally he wrote letters to his former employees, sometimes on technical matters, sometimes merely to give them fatherly lectures. In 1931 he queried Oscar Payne concerning Autorifle patents, adding:

> I have my son & his wife here with me, so I am not alone. They are expecting the birth of a child any day. I hope that it will be a boy. I know that Mrs. Payne & you love children. So do I. What a blessing they are as one grows older. . . . I was 70, Dec. 31, 1930 & have stopped *active* work. I hardly believe I will ever stop doing something; but dear friend, cultivate *play* & *good fun, recreation, games—even solitaire*, so you may drive away the ennui of old age—rather than be obliged to work—except in good causes & to make others happy![4]

In his later years Thompson began to take a greater interest in spiritual matters. Although nominally an Episcopalian, he had always been more philosophical than religious, and held to no church doctrines with any deep conviction. Some years earlier, during a late-evening conversation with Theodore Eickhoff, he remarked that he considered Christian faith largely a matter of "whether you want to bet on Christ or not."[5] As he grew older Thompson became more orthodox in his beliefs, but no less tolerant or philosophical. During an exchange of letters with Eickhoff at the time of George Harvey's death in 1928, Thompson expressed his fondness for Harvey and commented on other matters ranging from God to guns.

> Regarding the followers of the other great religious teachers than Christ (on whom I pin my faith—in Him & God more & more as time goes by), I cannot consider the merciful Lord as condemning those—not so fortunate as we—who lead lives in accordance with His laws.
>
> Our rifles so far have beaten the Garand & Pedersen in the tests, notwithstanding what you read—but we have not the resources to compete against our Government in business.
>
> Many thanks indeed for your information regarding the semi-dirigible airplane.[6]

Thompson's remarks about the rifle tests reflected his belief, shared by many other private inventors, that the Ordnance Department was not comparing its own designs and those of "outsiders" with complete impartiality. He suspected, too, that his former colleagues in the Ordnance Department bore him some resentment for having left the Army to develop arms privately. Whether or not Thompson was himself the victim of intentional Army prejudice would be hard to prove, but during the Twenties and early Thirties Colt officials came to similar conclusions—that private firms were disadvantaged—and several journalists of that period took the Ordnance Department to task for its policies. A newspaper editorial in 1931 charged that "The ordnance department is inordinately jealous of outside accomplishments and cherishes, in spite of experience, an obsession of superiority. . . . [The Department] seems determined to force its own device or go on experimenting until it gets one that is acceptable."[7]

Considering the headaches the gun had caused its inventors, the Army probably had sufficient reasons for rejecting the Thompson Autorifle with its inscrutable Blish lock. Later, the undeniable excellence of the Ordnance-designed Garand, proven in World War II, vindicated the Army's adoption of that rifle in 1936 instead of the privately developed Johnson rifle, whose supporters had angrily accused the Ordnance Department of prejudice. Doubtlessly the Autorifle was inferior to either. But in the case of the Thompson submachine gun, the Ordnance Department does seem to have tested unfairly and rejected for the wrong reasons. On the other hand, the Ordnance Department eventually adopted it for the wrong reasons, too, which suggests that the problem was not bias so much as backwardness.

When John Thompson first proposed a hand-held machine gun, and later when he demonstrated the working prototype, the Army reacted with enthusiasm. Military journals and military officials alike searched for new superlatives to describe its design, reliability, and firepower. This may well have been an old Army custom, for the same journals and officers seemed to have had praise for even the most dubious of new military inventions—including Larsen's iron cloud of an airplane with its armor plate and bellyful of Tommyguns. In any case, once called upon to officially adopt the submachine gun, the Army's feet grew cold. Judging from the way its tests were conducted and evaluated, the gun simply represented too radical a departure from tactical tradition.

In partial and preliminary tests during 1920 and 1921, the Thompson earned the usual accolades. In its final, official test in 1922, however, the Army found it to have little if any value as a combat weapon. This conclusion was virtually preordained. The Thompson was evaluated in competition with other types of small arms, and judged on the other weapons' terms.[8]

The tests were conducted at Fort Benning, Georgia, by the Department of Experiment, to determine the Thompson's accuracy, ballistics, reliability, and effectiveness in repelling night attacks. This last test seems particularly unimaginative in view of the submachine gun's applications in World War II, and probably reflects the Army's inability at that time to recognize the submachine gun's potential. It not only cast the gun in a purely defensive role, but envisioned it still as a trench weapon.

Not surprisingly, in its field trials the Thompson failed to greatly impress the testing board. The Springfield rifle surpassed it in single-shot accuracy; the Browning automatic rifle surpassed it in machine-fire accuracy at long ranges; and the sawed-off shotgun scored higher at very close ranges. In addition, the Army seems to have evaluated the scores rather arbitrarily. The Thompson was compared to the BAR, for example, by means of a mathematical formula which scored each weapon on the total points it earned for both firepower and accuracy at various ranges. These scores were then compiled in a table showing the theoretical number of Browning automatic rifles needed to equal the performance of one Thompson submachine gun at each range.

RANGE (Yards)	NUMBER OF BARS
50	2.66
100	0.98
150	3.73
200	4.36
250	8.66
300	5.41
350	3.43

400	1.10
450	0.157
500	1.00

From the above table a layman might judge the Thompson superior to the BAR at all ranges except 100 yards and 450 yards, and exactly equal to the BAR at 500 yards. The Army, however, interpreted the table as follows: "It is apparent from this table that only at very close ranges, where volume of fire is the dominant factor, is the Thompson gun superior to the Browning. Beyond 400 yards the advantage is to the Browning."[9]

While the Army found the Thompson's mechanical performance "satisfactory," it arrived at the following adverse conclusions:

- All .45-caliber pistol ammunition showed a pronounced break in accuracy between 200 and 300 yards.
- The falling off in accuracy in .45-caliber pistol ammunition between ranges of 100 and 500 yards is so rapid as to disqualify the ammunition at ranges over 300 yards.
- The trajectory of the Thompson submachine gun is so curved that the resulting danger space is insufficient and the probability of missing the target too great at battle ranges (600 yards).
- At pointblank range, the trench shotgun was superior to the Thompson submachine gun.
- Between 200 and 600 yards the Browning Automatic Rifle was superior to the Thompson submachine gun in fire against a screen 6 feet by 50 feet.
- Between 200 yards and 500 yards a single Springfield rifle was superior to the Thompson submachine gun in fire against silhouette figures.
- For night work at close range the Thompson submachine gun was little superior to the Browning Automatic Rifle.
- The rapid falling off in the accuracy of the Thompson submachine gun at ranges over 200 yards as the rate of fire is increased more than balances the increased volume of fire.
- Inaccuracy at ranges over 200 yards cannot be eliminated by expert shooting.
- The proper utilization of its large fire volume will necessitate a detail of ammunition bearers.
- The dominant quality of the gun is volume of fire, not accuracy.

- The stock, sights, drum magazine and carrying case are unsatisfactory.
- It is believed that no gun firing .45-caliber pistol ammunition can effectively compete with weapons of higher power, such as the Browning Automatic Rifle or the Springfield rifle.
- As an automatic pistol, the Thompson submachine gun does not possess sufficient advantages to warrant its being substituted for the present service automatic pistol.
- For special circumstances, such as trench warfare, the Thompson submachine gun has some advantages. It is believed that for similar conditions the trench shotgun is more valuable.[10]

Also, the board decided that the submachine gun "is in the class of a large automatic pistol, rather than in the class of a small automatic rifle." It simply did not occur to ordnance officials in 1922 to consider the submachine gun in a new class by itself, and their final recommendation was that "the Thompson Sub-Machine Gun be not adopted as infantry equipment."[11]

The Army gave the Tommygun no further consideration until the late 1920s, when the Cavalry modernized its tanks and developed armored cars. These new armored vehicles, the Army decided, needed some kind of light weapon of high firepower for use by their crews in case the vehicles were disabled in combat.

Thanks to constant nagging from Auto-Ordnance, the Army remembered the Thompson submachine gun. The Thompson met Cavalry requirements practically to the letter, but ordnance officials still seemed reluctant to adopt it and debated the matter for more than two years. Finally, in March 1932, the Army standardized the submachine gun as a "non-essential limited procurement" weapon for Cavalry use with armored vehicles "as a substitute for the semiautomatic rifle, development of which has not yet been completed."[12]

The Infantry remained unmoved, however. The commanding general of the Philippine Division, the Army's most likely candidate for the Tommygun, tested it and reported: "It is deemed inadvisable to add another weapon, still further complicating ammunition supply, to the already long list of Infantry weapons, this especially in view of the fact that jungle fighting will probably play only a minor role in operations of this command."[13]

A few years later, in 1936, the Cavalry decided that the Tommygun was in fact superior to the newly adopted Garand semiautomatic rifle by reason of its greater firepower and smaller size, and recommended its reclassification from "limited procurement" to "standard." In September 1938 the Thompson's reclassification became official, making it the Army's standard-issue Submachine gun, Caliber .45, M1928A1.[14]

But even in adopting the Thompson, the Army still failed to recognize its potential as an infantry weapon and issued it only to armored vehicles, scout cars, motorcycle troops, and machine-gun platoons, primarily for defensive use. Of the other major armies in the world, only the French and the British were equally hidebound in their tactical thinking. A British military writer of the same prewar period spoke for the Allies generally in his stingy predictions for the submachine gun's future:

> It is certainly true that the modern tendency is towards increasing the firepower of the infantryman, whether he still marches on his feet or whether he forms part of a motorized formation, but it is not in this direction that the value of the submachine gun lies; its true role will be to provide an emergency weapon for defense at close quarters for the personnel of specialist corps, such as motor transport drivers, spare numbers of machine gun units, tank crews, and headquarters details. In view of the enormously increased mobility and therefore strategic range of motorized formations it is likely that raids upon bases and communications will become relatively much more frequent, necessitating the use of the submachine gun as a strong defensive weapon against raiders.[15]

Raiders or other attacking elements would of course have no occasion to employ such a purely defensive weapon, or so Allied military planners then seemed to think.

Despite setbacks of every sort, the Thompsons never lost confidence in the submachine gun. Its poor reception in the early Twenties was understandable considering the times, and in 1929 it had seemed on the brink of success. In the Thirties, with Europe, Asia, and South America again in turmoil, and with the Army's grudging adoption of it, the Tommygun's prospects seemed extremely bright. Only now the Thompsons had no voice in Auto-Ordnance affairs, and Walter Ryan's firm policy was to not spend another penny of the Ryan fortune in a search for new customers. Both Thompsons firmly believed that at last the submachine gun was ready to earn enormous profits, but first they would have to regain control of the company.[16]

When he took over as president, general manager, and director of Auto-Ordnance in October 1930, Walter Ryan made it clear he wanted no further help from the Thompsons in running the business. From this time on, the only news they received of the company's activities was that contained in the meager minutes of annual corporate meetings. It soon became obvious, however, that Ryan had no intention of exploiting the expanding arms market, and that his sales policies could only result in the gradual dissipation of the company's physical inventory (its only real asset), with the proceeds going to the Ryan estate and into ordinary business expenses. This, of course, would eventually make Auto-Ordnance stock valueless to the Thompsons and other minority interests.

At the same time, the Ryan estate was trying to sell control of the company in a manner that would similarly dispossess the minority. The estate's plan was not to sell the majority stock, but to sell instead the notes of indebtedness, secured by chattel mortgage on the company's inventory of guns, with the majority stock thrown in as a bonus. The buyer would then foreclose on the mortgage and become sole owner of several thousand Tommyguns, effectively liquidating the minority stockholders' interest.

Several times Marcellus Thompson and his attorney, Thomas A. Kane of New York, blocked such sales by threatening or starting court actions. Several other times Marcellus proposed new plans for refinancing the company, believing he could liquidate the existing inventory in a manner that would develop sales, create a market, and justify putting the Tommygun back into production—this time to everyone's profit. To substantiate his claims, Marcellus embarked on an unofficial one-man sales campaign and for two or three years hounded the Guaranty Trust Company with letters from a dozen different countries evincing interest in the gun. The executors, however, were never much impressed, nor would they grant Thompson an exclusive agency to sell the gun. Each time Thompson developed an important sales prospect, other agents, claiming closer ties with Auto-Ordnance, would intervene with competing offers that invariably led to a breakdown of negotiations. To Thompson, such interference was a calculated effort by the Ryans to sabotage sales that would have helped the minority stockholders secure outside financial support to regain control of the company.

With no fat sales or manufacturing contract to wave in front of a prospective financier, Thompson found money-raising an all but hopeless task. In person or by letter he contacted well over a hundred investment and manufacturing firms, including Remington, DuPont, Sperry Gyroscope, Chrysler, Winchester, even the National Fireworks Company. No one, however, was even mildly interested in putting up the substantial amount of hard cash it obviously would take to buy out the Ryan interest in Auto-Ordnance. A warehouseful of old Tommyguns and Marcellus Thompson's boundless optimism were not enough security.

For eight years the Ryans fiddled and the Thompsons burned, watching their company's assets trickle away one small order at a time. By the end of 1938 Auto-Ordnance had sold a total of 10,300 submachine guns—4,700 in the United States (mostly to police, penal institutions, protective agencies, state militias, and industrial guards), 4,100 in foreign countries, and 1,500 to the United States government. Gross income on guns, parts, and accessories since 1921 totaled $2,234,000, most of which had gone into development, operating, and sales expenses. Business losses (including the interest accruing on debts owed the Ryan estate) for the years 1936, 1937, and 1938 were $82,200, $62,840, and $91,440 respectively, leaving Auto-Ordnance in debt to the Ryan estate some $1,090,000, plus accrued interest amounting to $1,216,661. The company's total assets amounted to $412,300, almost all of which was represented by the 4,700 Thompson submachine guns still left in inventory.[17]

In his long struggle to promote sales and raise capital Marcellus Thompson had virtually exhausted his personal savings, only to find himself no better off than when he started. If anything, his situation seemed more hopeless than ever. Then, one day in the middle of January 1939, opportunity suddenly knocked—presenting itself in the person of one Matthew J. Hall, a Wall Street broker who for years had been trying to sell Auto-Ordnance on behalf of the Ryan estate, on terms that the minority stockholders had vigorously opposed.

As Hall explained it, the Ryan heirs and executors both were tired of hanging onto Auto-Ordnance, which by this time, 1939, represented the only holding of the Ryan estate not yet liquidated. The time was ripe, therefore, to make the Ryan family a tempting cash offer for their controlling interest in the company. Hall went on to suggest that the cash be raised through the public sale of stock, and he just happened to know a financier who might be willing to underwrite the issue.

The underwriter Hall had in mind was Russell Maguire, a forty-two-year-old Connecticut industrialist and promoter with offices at 1 Wall Street.[18] Born in Meriden, Connecticut, in 1897, Maguire briefly studied electrical engineering at M.I.T. and commercial law at N.Y.U., but found both fields too dull for his tastes. After serving as a Navy flier in the First World War, he took a job as a broker's clerk and discovered finance, which looked to be not only stimulating but highly profitable. In 1920 he went to work for a casualty company in Philadelphia, and soon afterward opened his own business as a surety bond agent and real estate speculator. In 1925 he organized the Penn Company, a guaranteed mortgage firm, and did well enough until realty values crashed along with the stock market four years later. The company was liquidated, and Maguire went into personal bankruptcy with liabilities of $393,133 and assets of $169.

Discharged from bankruptcy in the spring of 1933, Maguire abandoned his old creditors and went to New York in search of new capital. In 1935 he accumulated enough to organize a securities firm, Russell Maguire & Company, and became an aggressive and fairly successful underwriter of industrial stock issues. Later he opened a second securities firm, Maguire & Company, in Delaware.

Maguire began his comeback by buying out the nearly defunct Alco Valve Company of St. Louis and boosting its earnings from

$21,000 to $900,000 in a single year. He had equally good luck with several other insolvent companies and began to specialize in salvaging neglected firms. Everything he knew about finance and management, Maguire was fond of saying, he had learned in the business world's "school of hard knocks." He had a good "nose for a deal," and the intuitive ability to "smell out a guy who knows what he's talking about."[19] He also had a weakness for clichés.

With his talented nose and a little help from Matthew Hall, Maguire smelled out the Auto-Ordnance Corporation and found it adequately neglected. With three or four employees, a static inventory of twenty-year-old guns, and an enormous debt, it was virtually moribund. Moreover, he reckoned that Marcellus Thompson knew what he was talking about when he predicted that Europe was headed fast in the direction of a major war that would create an enormous market for submachine guns. With little further discussion, Maguire agreed to underwrite a public stock issue if Thompson could get the Ryans to bind themselves with a contract of sale.

About the first of February 1939, Marcellus Thompson and Thomas Kane took their proposal to the Guaranty Trust Company. The Ryan executors were agreeable to selling, but somewhat skeptical. Thompson had earned himself a reputation for concocting unrealistic financial schemes to buy out the Ryan interest, and over the years, had made himself a considerable nuisance to the hard-nosed bankers in Guaranty's trust department. The executors would sell, they explained, because the Ryan heirs were pressing them to dispose of the property; and they would sell at a close-out price of $529,000. In addition, they would write off the $1,090,000 in notes still owed the Ryan estate, as well as the accrued interest amounting to $1,216,661. They made only two stipulations: the sale price would be paid in cash, and in the event the deal fell through, for any reason, the minority stockholders would permit the Guaranty Trust Company to liquidate the company in any manner it desired. To guarantee this, the executors wanted a special clause in the sales contract. The minority stockholders would deposit all of their stock with Guaranty Trust as liquidated damages in the event of default or failure to meet the terms of the sales agreement.

Such a provision put the Thompsons and the heirs of Commander

Blish far out on a fragile limb. If Maguire failed to keep his end of the bargain, the money end, the minority interests would be wiped out, and the Ryans would be free to do as they pleased with Auto-Ordnance. But the Thompsons and the Blish family felt they had no better choice. After some further weeks of bargaining, Kane, Guaranty Trust, and Russell Maguire worked out the details of a formal sales agreement. It required the payment of $529,000 in two equal cash installments of $264,500, the first due on July 21, 1939, the second due on September 21, 1939. Upon payment of the total amount, the Ryan estate would transfer to the purchasers its entire interest in the Auto-Ordnance Corporation, consisting of 18,505 shares of capital stock, plus all notes of indebtedness and accrued interest. In the event the purchasers made the first payment but not the second, the first payment would be retained by the Ryan estate as further liquidated damages in addition to the minority stock deposited with Guaranty Trust.

To raise the necessary capital, Thompson and Maguire formed the Thompson Automatic Arms Corporation, in March 1939, and entered into an underwriting agreement. The minority stockholders would exchange their Auto-Ordnance shares for shares in Thompson Automatic Arms, and Maguire & Company agreed to sell 300,000 shares of Thompson Automatic at $2.75 a share, to net $2 a share. This would furnish the new company with capital of $600,000 with which to buy the Ryan interest in Auto-Ordnance, leaving a difference of $71,000 to serve as working reserve.

This plan was fairly simple and straightforward, but ran into trouble almost immediately. On May 23, 1939, Thomas Kane filed the new company's stock registration statement with the Securities and Exchange Commission, requesting the commission to approve the issue as quickly as possible so the stock could be sold before the Guaranty Trust deadline. When nearly a month had gone by, Kane began to worry. Several times the commission had requested more information, indicating it was having doubts about the stock. If the SEC ultimately refused to authorize the issue, all deals were off.

Kane immediately alerted Thompson and Maguire to this looming possibility, urging them to find an alternative source of capital without delay. On June 27, with the deadline less than a month away, he went to the SEC to find out the status of the registration application. The news was totally bad. The commissioners had not yet made a final decision, but virtually assured him that the SEC was not going to authorize such a speculative stock issue. In the commission's view, Auto-Ordnance had no assets of any value—only obsolete guns, expired patents, and some few pieces of worn-out machinery. It had no market, foreign or domestic, and no worthwhile wars in progress or in prospect. Clearly, the Tommygun's future was as bleak as its past, and Thompson Automatic Arms merely a scheme to bail out Auto-Ordnance stockholders by bringing in the public. Nor was Marcellus Thompson able to persuade the commission otherwise at a conference the next day.

On July 3 a new development took place. With no warning whatever Auto-Ordnance received an order from the United States Army for the immediate delivery of nine hundred submachine guns, plus parts and accessories, at a price of $454,000.

Surprised and delighted, Thompson rushed back to the SEC waving a photostatic copy of the Army's order. The SEC examined the order and listened patiently to Thompson's assurances of another world war, but was not impressed. The order, the commissioners decided, was simply a windfall.

Turned down by the SEC, Thompson pleaded with Guaranty Trust to extend the contract deadline, which was less than three weeks away. The executors, however, would not budge. The Army's submachine gun order had not swayed the SEC, but it thoroughly impressed the Ryan heirs, who realized suddenly that they were giving away the entire company for little more than the Army's paying price for one order of guns. If Thompson could not raise the money by July 21, the Ryans only had reason to rejoice.

Caught between the Ryans and the SEC, Thompson tried to impress Russell Maguire with the urgency of the situation. Maguire assured him that he would attend to the matter. He said nothing, however, to indicate he had any particular plan in mind, and in fact seemed, at least to Thompson, quite disinterested in the problem.

Losing confidence in Maguire, Thompson and Kane desperately set out in a new search for capital, this time armed with evidence that the Tommygun was a potential money-maker. By July 17 they had found a likely source of funds in the Marine Midland Bank & Trust Company, only to discover that Maguire was on the same track and was also arranging a loan. In a conference among all parties concerned, Marine Midland expressed its intentions of granting the loan, but only to one group. Maguire quickly pointed out that his underwriting agreement still was in force. Reluctantly, Thompson and Kane withdrew their loan application and stepped aside.

On July 19, Maguire, now thoroughly in control of the situation, began letting the Thompson interests in on things. He had arranged for a loan of $539,000 by pledging all assets of Auto-Ordnance, plus personal holdings worth $100,000; the loan would be ready at 2 P.M. July 21, deadline day, and would be paid immediately to Guaranty Trust in a lump sum, with $10,000 left over for initial operating expenses. Finally, as payment for his services in obtaining the loan, including the pledging of personal property, Maguire demanded that he be voted a block of 116,400 shares of stock in the new company.

Maguire's price for services rendered came as an unpleasant surprise to Thompson and Kane, but the immediacy of the situation left no time for haggling. A board of directors meeting was called for later the same evening and the stock was voted to Maguire. In addition, Thompson and Kane were each voted 1,800 shares for their own services, bringing their personal holdings up to 51,107 and 14,743 shares respectively. At the same meeting Thompson was elected president of the corporation, and Kane vice president and treasurer, with power to act in the president's absence.

These actions were taken on the Wednesday evening before the Friday, July 21, deadline. Thursday was a day of frenzied effort by all parties to iron out last-minute details and get all papers in order. By late that night the work still was not finished, and after a few hours sleep Thompson and Kane went to the office of Maguire's attorneys early Friday morning to try to wind things up before meeting with Guaranty Trust that afternoon. When he arrived at the office, Thompson felt and looked exhausted, but attributed it to lack of sleep and the strain he had been under in the last month. About 10 A.M., as he worked over a large pile of papers, he suddenly collapsed in great pain.

Thomas Kane rode in the ambulance that took Thompson to the Broad Street Hospital, saw to his admission, then rushed back to the attorney's office, where he felt Marcellus needed him most. He still had not finished the legal documents when an officer of Guaranty Trust called to tell him that the Ryan trustees were assembled and waiting to close the deal. Kane asked for more time, explaining that Marcellus Thompson had just suffered a severe stroke. He was told to come as soon as possible, as the deadline would not be extended under any circumstances. Unless the closing took place that day, the minority stockholders would be declared in default and their holdings forfeited.

With time running out, Kane rushed back to the pile of records and documents, only to be interrupted a few minutes later by one of Russell Maguire's business associates. Maguire, the man explained, would require further compensation for his services—another 7,500 shares of stock, transferred to him from Thompson (4,900 shares) and from Kane (2,600 shares). Nor would Maguire close the deal with the Ryans until he received the stock.

Kane was aghast. Such a transfer would give Maguire 50.8 per cent of the issued stock and a controlling interest in the corporation, and the scoundrel was springing this demand in a zero-hour ultimatum that Kane and Thompson could scarcely afford to reject. Nonetheless, Kane rejected it. Stopping work on the papers, he angrily informed his visitor that Marcellus was in the hospital critically ill and completely incapacitated, and that he himself did not intend to yield to such an unscrupulous attempt at extortion.

In response to Kane's rebellion, Maguire sent over one emissary after another to reiterate the demand and urge Kane to reconsider. In his own view, Maguire was swindling no one. He had agreed initially only to underwrite a stock issue, not gamble everything he owned on someone else's venture, and he felt entitled to substantial compensation. Moreover, neither Maguire nor his business associates had any confidence whatever in the Thompson group's ability to administer the new company. Marcellus was too obviously an amateur and a visionary, and since he would not be losing his original interest in the company, Maguire was only doing him a favor by taking over the controls. Most of all, Maguire did not want a bunch of incompetents managing his personal fortune.[20]

But to Kane it was still extortion, and he refused to budge, which put Maguire in something of a bind. The Ryan estate's trustees and the representatives of Marine Midland had been assembled in the board room of Guaranty Trust since 2 P.M. and were getting restless at the delay, which Maguire blamed on unfinished paperwork. At five o'clock the Ryan attorneys started picking up their papers to leave. Maguire quickly reminded them that the deadline was midnight, not five o'clock, and insisted on meeting again at eight. Meanwhile he would apply a little more reason to Kane, who was stubbornly holding his ground in the office of Maguire's attorney, refusing to sign over any stock or even do any more work on the documents.

Shortly before eight o'clock Kane received a visit from Matthew Hall, who had brought Maguire and Thompson together originally. Hall had a paper bearing Marcellus Thompson's signature, which he had somehow obtained in Thompson's hospital room. It was an agreement to transfer 4,900 shares of stock to Maguire as soon as he was physically able.

This document placed on Kane the full responsibility for any further delay in the closing. Still, he held out, hoping Maguire would relent. At nine o'clock another call came from the board room at Guaranty Trust: unless Kane came over immediately, a default would be declared, and the Ryan trustees would take Auto-Ordnance and go home.

This left Kane no choice but to pack up his papers and go. When he arrived at the board room about 9:30, Maguire approached him personally, repeating his demands and threatening in no uncertain terms to wreck the entire deal if Kane balked any longer. He reminded Kane of his responsibilities to others: to Thompson, whose illness was probably terminal, and who had no other assets to leave a widow and two minor children; and to Ida Blish, widow of John Blish, who likewise had no other assets, and who at that moment waited in a New York hotel room confident that Kane would protect her interests. Finally, Maguire pointed out, forfeiture of the minority stock would only free him to reopen negotiations with the Ryans on Monday, this time with no interference from Kane, the Thompsons, Mrs. Blish, or anyone else. Deciding that Maguire fully intended to carry out his threats, and with the deadline now less than two hours away, Kane surrendered.

A little after 10 P.M. Kane signed the papers transferring 2,600 shares of stock to Maguire, and also executed resignations of office for himself and Marcellus Thompson, to take effect upon Maguire's request. Then he returned to work on the papers, racing against the board room clock. The final documents were signed fifteen minutes before midnight.

Reborn as a wholly owned subsidiary of Thompson Automatic Arms, Russell Maguire president and director, Auto-Ordnance immediately bustled with activity. Maguire completely reorganized the company, launched a promotional campaign, and optimistically placed a large order for gun steel. But then his troubles began. If Marcellus Thompson knew little about big business, Russell Maguire knew little about the gun business, and soon discovered that some major obstacles stood in the way of his making any quick financial killing with his newly acquired machine gun.[21]

For one thing, the Army proved to be in no hurry to order any more Thompsons. For another, the huge foreign market Marcellus had pictured did indeed want submachine guns, but chambered for foreign service ammunition, especially 9mm. When Germany invaded Poland on September 1, setting off the world war Thompson had predicted, the Tommygun's prospects only faded, as the Neutrality Act of 1935 automatically imposed an embargo on all military supplies to the warring countries. Nor could Maguire get the gun into production when the act was revised in November 1939. After ordering his gun steel, Maguire began shopping for a subcontracting manufacturer, only to learn that none of the country's armament firms were the least bit interested. Colt's wanted nothing more to do with the Thompson. Since 1921 the company had been blamed too many times for the misdeeds of Tommyguns marked "Manufactured by Colt's Patent Fire Arms Mfg. Co., Hartford, Conn., U.S.A." Anyway, Colt's was busy making Browning machine guns for the Army. The other major arms makers were doing nicely with civilian and sporting arms and had no desire to tool up for a foreign war that might either fizzle out or embroil them in another "iron, blood, and profits" scandal.

For three months it appeared that the new Auto-Ordnance Corporation might do no better than the old one, but the outlook soon brightened. On November 1 the French purchasing commission ordered 3,000 guns at a contract price of $750,000. About the same time the British purchasing commission made inquiries that indicated Britain's intentions of placing substantial orders in the near future.

With the Tommygun finally showing potential, Maguire went to the Savage Arms Corporation, then in Utica, New York. Savage officials had turned him down once already, but not as firmly as they might have. This time, with one sale in the bag and another on the line, Maguire induced the company to subcontract for the manufacture of 10,000 Thompsons, Model of 1928.

Savage undertook the job with something less than enthusiasm. Auto-Ordnance would have to provide or pay for all manufacturing equipment, matériel, and inspection personnel; would have to deposit a 50 per cent down payment before work began on the order; and then would have to pay the full contract price per gun as the guns came off the assembly line. Under this curious arrangement, the entire order would be paid for in full when only half of the guns were delivered.[22] Savage was taking no chances.

Auto-Ordnance contracted with Savage on December 15, 1939, and immediately shipped to Utica the machinery that had been stored at Colt's since 1922. Except for the cutting tools, which were 40 per cent worn, the equipment was in relatively good condition. Savage refurbished it, installed it, added to it, and began delivering finished guns in April 1940.

By the time Savage got the Tommygun into production the war's complexion had changed radically, creating an enormous demand for submachine guns of any kind. Between February 1940 and the end of the year, the British placed thirteen orders for 107,500 Thompsons at an aggregate contract price of $21,502,758. In March the French ordered an additional 3,000 guns, and by December the United States Army had contracted for the manufacture of 20,450.[23]

Dunkirk had made the difference. After that catastrophe the British, by transatlantic phone, gave Tommyguns a priority second only to that of airplanes, and so did the Russians.[24] By the end of 1940 orders were piling up into the hundreds of thousands as every Allied country, including the United States, clamored for Thompson submachine guns.

For almost twenty years the Tommygun had scandalized its inventors and distressed its investors. Then, almost over night, it turned into a vital, highly profitable piece of military ordnance. Ironically, neither John Thompson nor Marcellus lived quite long enough to benefit from their invention, or even to know the important role it would play in the Second World War. On October 17, 1939, three months after his stroke, Marcellus died in the Broad Street Hospital at the age of fifty-six. Eight months later, on June 21, 1940, John Thompson died of a heart attack at his home in Great Neck, Long Island. He was seventy-nine.

The last few years of Thompson's life were difficult ones. Crippled with arthritis, eyesight failing, he spent most of his time in bed or in his wheelchair. He was cared for still by Willy Smith, his onetime orderly, later his butler, and finally his last close friend and companion. The two old men reminisced often on their many years together, and entertained occasional guests with their "war stories" and playful exhibitions of their soldierly skills. From his wheelchair, General Thompson would "order the troops to fall in," then drill his one-man army around the living room.[25]

Despite his age and poor health, Thompson remained alert, well informed, cheerful, and somewhat vain about his appearance. In 1939 *Time* published his photograph with an article on the new Thompson Automatic Arms Corporation. It was a clear, rather flattering picture of an aging general with white hair and a carefully trimmed beard and mustache, looking fit and quite pleased, but Thompson thought it dreadful. He sent a clipping of the article to his old friend and onetime neighbor, Arthur Krock, the *New York Times* columnist, after first removing the picture. In the margin of the clipping he explained, "Taken by flashlight in my dark room on a dark, rainy, heavy day, was so impossible I cut it out.—J.T.T."[26]

The submachine gun remained very much in Thompson's thoughts at all times, a source of mixed feelings. It pleased Thompson greatly when the Army finally adopted the submachine gun as Cavalry standard in 1938, but he passed the news along to Theodore Eickhoff with a touch of sarcasm. "The T.S.M.G. has *at last* been adopted by the War Dept. as a 'Service Weapon' for 'short range combat.' At that range, 300 yards—the critical moment of a combat—the band and other noncombatants should discard their instruments and rush in!"[27] In another letter, for no apparent reason and without elaborating, he referred to his long-cherished Blish patent in words seeming to hint that he had once discovered and thereafter concealed what he considered an embarrassing fact.

> I found that the Blish principle was really invented by a French Artillery Officer. This does not bother me. I went ahead with the spending of over one million dollars in the development of the Thompson Sub Machine Gun. I did not believe any other group had the necessary confidence, knowledge, and skill to develop it. Again I am so grateful to you for all this. . . .
>
> If I do not promptly answer your letters please know that I am physically unable to do so. I still have considerable work to do in keeping up my little family, whom I am glad to have with me.[28]

This letter, written July 27, 1939, made no mention of his son's stroke a few days earlier, although by then it was apparent that the stroke would be fatal. When Marcellus died eleven weeks later, Thompson took the news stoically, but thereafter his health and spirits declined rapidly. Through Marcellus he had been able to participate, at least vicariously, in all of the planning and struggling to revive the submachine gun. After his son's death, in which the gun too clearly played some part, Thompson stopped writing to his old friends, and for the first time in his life, stopped planning new projects for the future.

Eickhoff heard from Thompson for the last time shortly before Marcellus died. It was a depressed, melancholy letter in parts, written almost illegibly in pencil on the covers of a small booklet promoting the Thompson submachine gun. It ended on a cheerful note, nonetheless, with the customary words of fatherly advice:

> I do not thoroughly like, of course, the Ordnance semi-automatic rifle [Garand] because the gas passes through a very small hole—likely to stop fire with our raw war troops! So I do not think it will prove as good a rifle as the sure acting Thompson automatic rifle cal. .30 which you & Payne did so much to develop. I share all honors with you both & Goll. . . . Your scientific training and acumen helped me in our preliminary experiments . . . with the old pioneer Blish principle at Chester, when you were a welcome member of my family in my home at Chester, Penna. I often think of you and our happy old days at Chester and Washington, D.C.
>
> I have given *my valedictory* to arms, as I want to pay more attention now to saving human life than destroying it. May the deadly T.S.M.G. always "speak for" God & Country. It has worried me that the gun has been so *stolen* by evil men & used for purposes outside our motto, "On the side of law & order". . . .
>
> I have been quite sick & confined to my room & bed lately—the result of hard work all my life—but I am happy & keep a peaceful mind. Don't worry, and keep a peaceful mind and stomach. . . .
>
> Your grateful & devoted old friend,
> John Taliaferro Thompson[29]

9

World War II

The deadliest weapon, pound for pound, ever devised by man.

—*Time* magazine, *1939*

The sudden worldwide demand for Tommyguns did not take Maguire by surprise. To the contrary, he had gambled heavily on it, and he continued to gamble that the demand would increase enormously. With foresight and at considerable risk he built a giant manufacturing enterprise geared to market conditions and materials shortages that did not yet exist.[1]

Even before the Savage plant at Utica went into production, Maguire was buying more steel, ordering machine tools, and shopping for a site to open his own factory. In August 1940, Auto-Ordnance leased (and later bought) an old brake-lining plant consisting of an entire block of buildings along Railroad Avenue in Bridgeport, Connecticut. After renovating the buildings, the company launched a crash program to get the factory into production and to make it both self-contained and self-sufficient. Rather than wait months for the delivery of new machine tools, Auto-Ordnance purchased mostly secondhand equipment, which it rebuilt or overhauled at the time of installation. Some of the machinery was antique. Several milling and profiling machines dated back to World War I, and one Pratt & Whitney profiler had turned out rifle parts during the Civil War. But even the oldest machines, once rejuvenated and given independent motor drive, worked as well as any.

In January 1941 the engineering department moved into the renovated buildings and began laying the groundwork for mass production. Anticipating a shortage of cutting tools and gauges, the company first set up facilities to manufacture all of its own tools, fixtures, jigs, cutters, and gauges, and thereby make itself independent of the national machine-tool market.

During its first six months of operation the Bridgeport plant manufactured nothing but tools and gauges. Next, it began turning out the submachine gun's larger components, such as frame and receiver, subcontracting with other companies for barrels, stocks, and most of the internal parts. The first Bridgeport-built Tommyguns were completed on this basis in August 1941. Gradually the company expanded its manufacturing operations to become an entirely self-contained factory that turned out finished, tested guns from foundry materials.

Russell Maguire had little in common with John Thompson, but like Thompson he placed great importance on the human factor in the manufacturing business. "I may not know how to put a plug in a wall," Maguire would say, "but I know how to put people together and get them started. I chart a course, pick the men, and give authority with responsibility."[2]

The men Maguire picked were not gunmakers by profession, and they felt no obligations to the traditional manufacturing techniques of the old New England small-arms industry. Having no preconceived notions about gunmaking, they found numerous manufacturing short cuts. Metal stamping and punching operations replaced some of the work usually done on milling and profiling machines. Hand filing was reduced to a minimum through the use of special power tools, and by devising a new way to remove burrs and sharp edges by "tumbling" small parts in barrels. When it proved impossible to achieve a balance in production facilities using a line production system, Auto-Ordnance shifted over to a departmentalized system that kept all machines busy and all parts coming at the same rate.

In June 1939 the Army placed its first sizable order, for 950 Thompsons. By the end of 1940 the Army had placed additional orders totaling 20,450 guns, and by the end of 1941 this figure had been raised to 319,000, including orders from Allied countries through the Lend-Lease program. The Auto-Ordnance plant at the Savage company in Utica went into production in the spring of 1940 and by February 1941 was turning out finished guns at the rate of about ten thousand a month. By June this rate had doubled, and in October, the month that the Bridgeport plant came fully into production, Auto-Ordnance was able to deliver 42,930 guns. In February 1942 the company manufactured its five hundred thousandth gun, and by the summer of that year had achieved a combined production rate at both plants of about ninety thousand guns a month. From 1940 through 1944, when Thompson production ended, Auto-Ordnance manufactured a total of 1,750,000 submachine guns, plus spare parts equivalent to 250,000 more.[3] Most of the guns, about 1,250,000, were built at the Utica plant and bear an "S" (for Savage) preceding the serial number. Those manufactured at the Bridgeport plant have the serial prefix "A.O."

The company's phenomenal production rate earned Auto-Ordnance five Army-Navy "E" awards for outstanding defense work. It received the first of these, for small arms production, in September 1942. The presentation ceremony was an affair held in November in a huge circus tent decorated with Tommyguns and a greatly enlarged *Saturday Evening Post* cover picturing a soldier holding a Thompson. War correspondent Quentin Reynolds served as master of ceremonies, and Wendell Willkie attended as guest of honor. The guest speakers, including military representatives of the United States, Britain, and China, praised the Thompson submachine gun and lauded the company's production record. Britain's Lord Mountbatten

The St. Valentine's Day Massacre in Chicago, 1929. (UPI)

George "Machine Gun" Kelly being escorted from the Shelby County jail in Memphis in 1933 by Federal agents armed with Model 1928 Thompsons. (UPI)

In December 1933 Chicago police looking for Dillinger killed three other wanted criminals in a machine-gun battle in a Rogers Park apartment building. (UPI)

John Dillinger posing with a Thompson and the wooden pistol he claimed he used in breaking out of the Crown Point, Indiana, jail in 1934. (Wide World)

Arsenal of submachine guns and other weapons found in the Florida hideout of "Ma" and Freddie Barker after they were killed in a machine-gun battle with Federal agents in January 1935. (FBI)

Bullet- and bomb-proof limousine of Al Capone, complete with Tommyguns, was exhibited throughout the United States and England during the early 1930s. (UPI)

Baltimore detectives James Holden and J. E. Farrell inspect Tommyguns and other weapons found in the apartment of Basil "The Owl" Banghart, suspected of kidnapping Chicago underworld figure Jake "The Barber" Factor in 1933. (UPI)

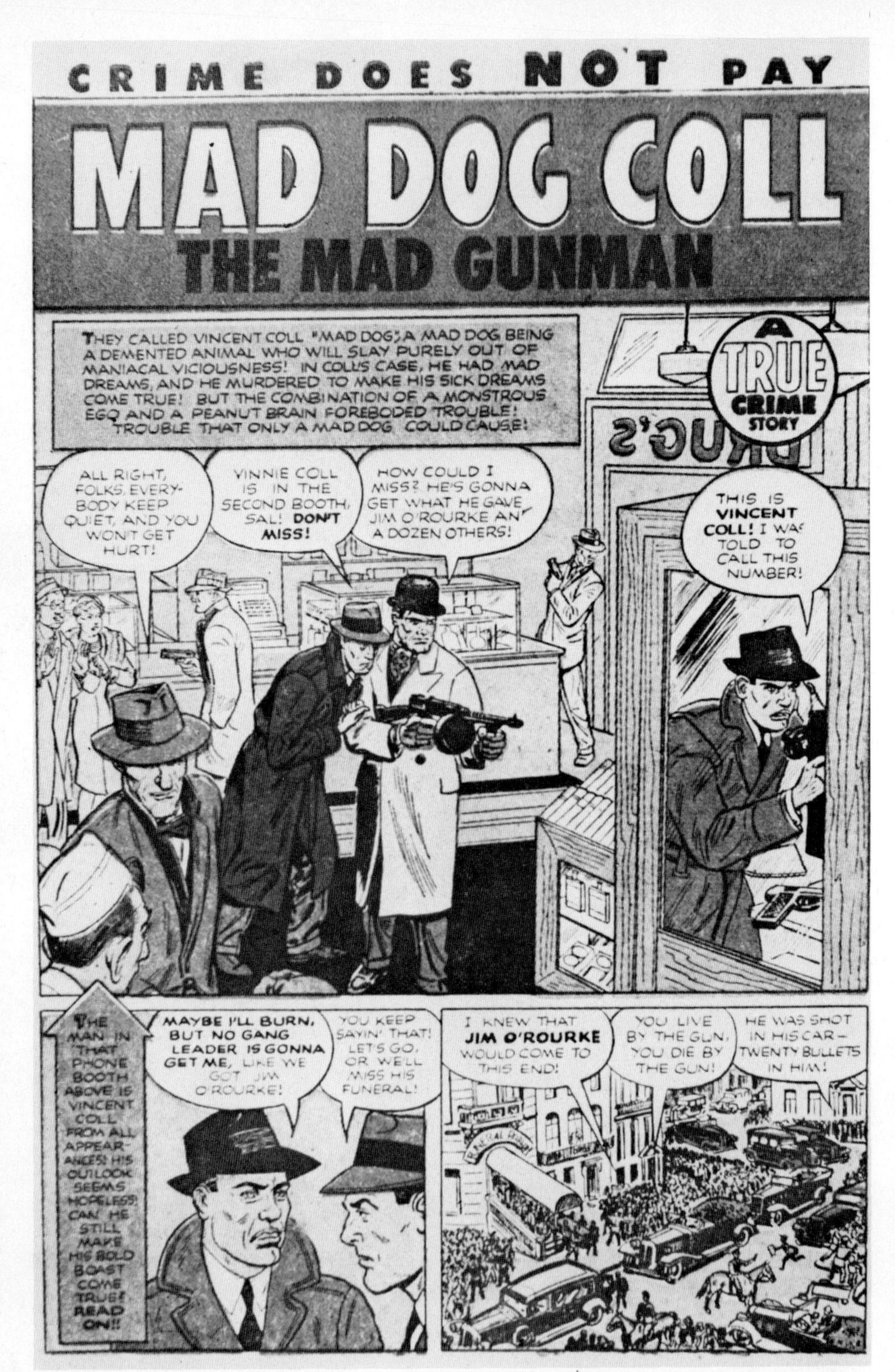

Page from an old *Crime Does Not Pay* comic book portraying New York gangster "Mad Dog" Coll, who was machine-gunned by the Dutch Schultz mob in a New York drugstore telephone booth in 1932.

Two Marine Corps sergeants named Wilson (left) and Robinson after a fight with Nicaraguan rebels near La Palancer in the summer of 1928. (USMC)

U.S. Marines patrolling the streets of Shanghai in 1932. (USMC)

Marine machine-gun post guarding the U.S. Legation in Shanghai in 1927. (USMC)

Model 1927 Thompson Semi-automatic Carbine, identical in appearance with the submachine gun. (George Goll)

Model 1929 B.S.A.-Thompson submachine guns chambered for 9mm (above) and .45 ACP (below). (George Goll)

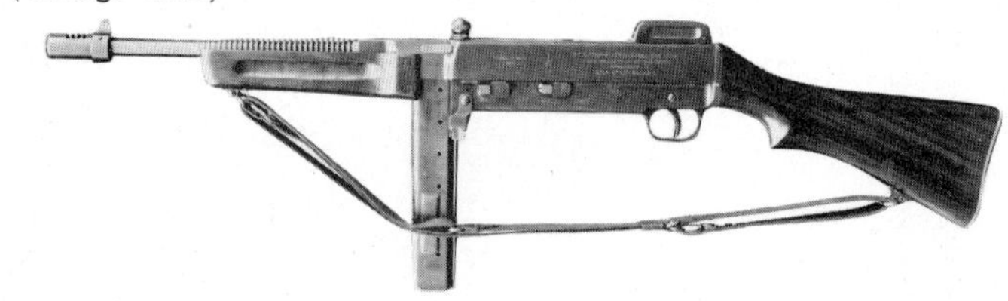

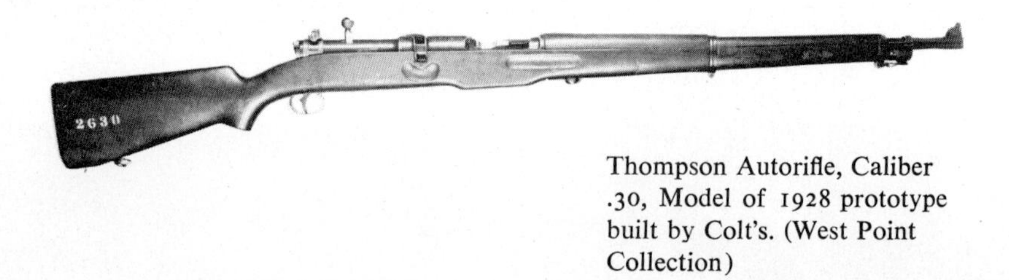

Thompson Autorifle, Caliber .30, Model of 1928 prototype built by Colt's. (West Point Collection)

Thompson submachine gun, U.S. Navy Model of 1928, with Cutts compensator. (George Goll)

George Goll demonstrating a Thompson submachine gun for the Connecticut State Police in 1938. (George Goll)

Young Irish Republican Army "guerrillas" photographed in 1957 with their weapons, including two Thompsons probably smuggled into Ireland during the 1920s. (UPI)

Cumberland, Maryland, police armed with Thompsons covering the arrival of the Bonus Army on its march to Washington in 1932. (UPI)

Submachine gun developed by George Hyde to compete with the Thompson. (U.S. Army)

Snapshot of John Thompson taken in New Canaan about 1930 and sent to Theodore Eickhoff at Christmas, 1939.

"The Hedges," John Thompson's home on Bank Street in New Canaan, Connecticut, where he lived from 1919 until his wife's death in 1930. (T. H. Eickhoff)

U.S. Cavalry officers examine Thompson-equipped motorcycle in 1938. (George Goll)

U.S. Army motorcycle couriers with Tommyguns, 1943. (UPI)

U.S. Army Air Corps security squad armed with newly issued Thompsons at Hamilton Field, California, in October 1941. (Wide World)

British Prime Minister Winston Churchill examining American Tommyguns while on a tour of British defense installations in July 1940. (Imperial War Museum, London)

Winston Churchill. (Imperial War Museum, London)

Russell Maguire, about 1944. (Wide World)

Auto-Ordnance factory at Bridgeport, Connecticut. (Raymond Koontz)

Russell Maguire presenting a scale-model Tommygun to Evelyn Thompson, Marcellus Thompson's widow, in 1940. (Thomas Kane)

Russell Maguire addressing Auto-Ordnance employees and Allied government representatives at an Army-Navy "E" award ceremony in 1942. (Raymond Koontz)

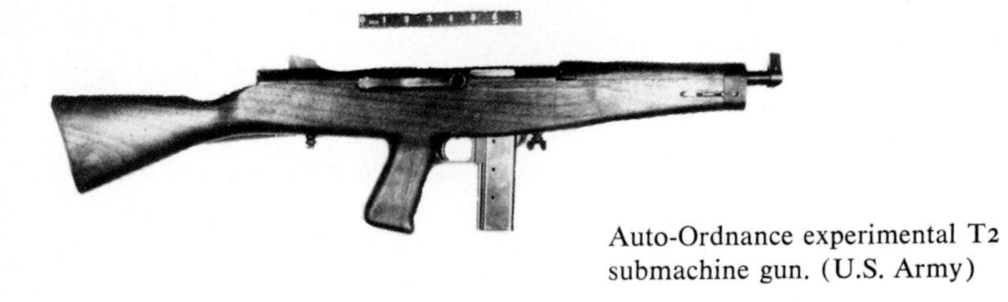

Auto-Ordnance experimental T2 submachine gun. (U.S. Army)

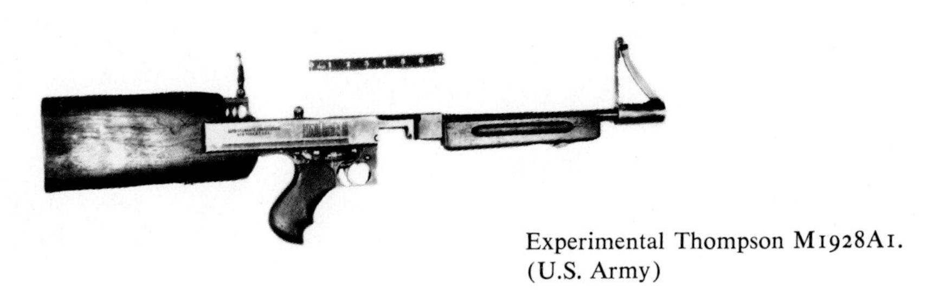

Experimental Thompson M1928A1. (U.S. Army)

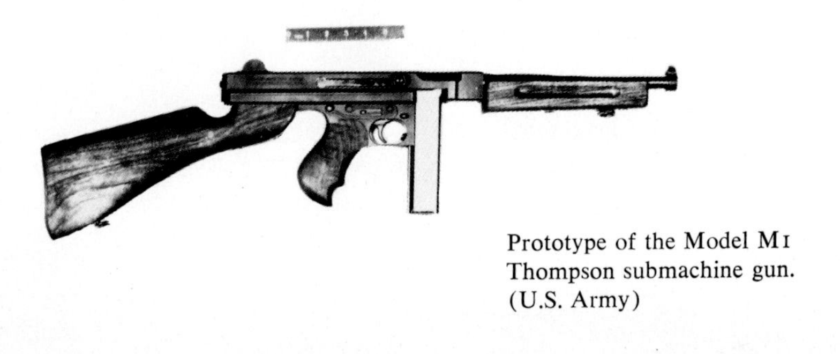

Prototype of the Model M1 Thompson submachine gun. (U.S. Army)

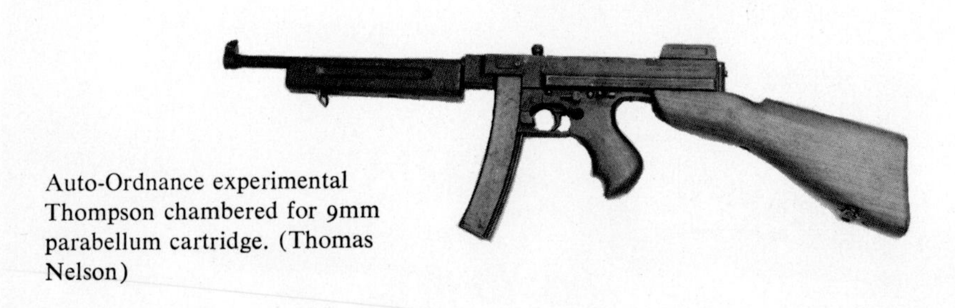

Auto-Ordnance experimental Thompson chambered for 9mm parabellum cartridge. (Thomas Nelson)

British STEN submachine gun.
(U.S. Army)

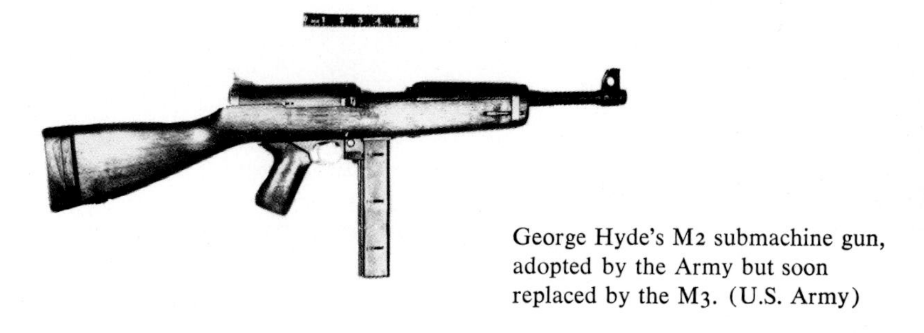

George Hyde's M2 submachine gun, adopted by the Army but soon replaced by the M3. (U.S. Army)

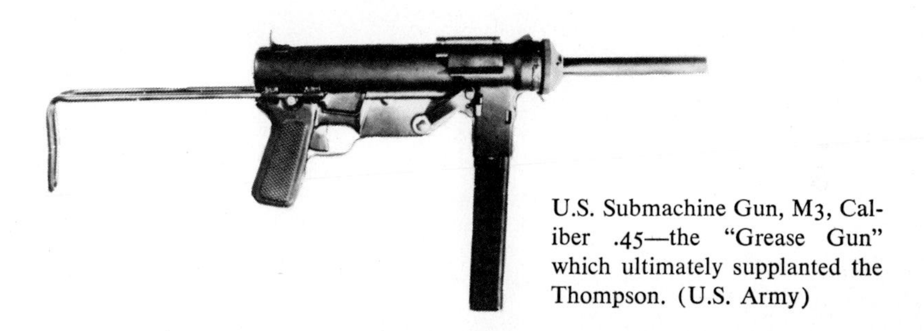

U.S. Submachine Gun, M3, Caliber .45—the "Grease Gun" which ultimately supplanted the Thompson. (U.S. Army)

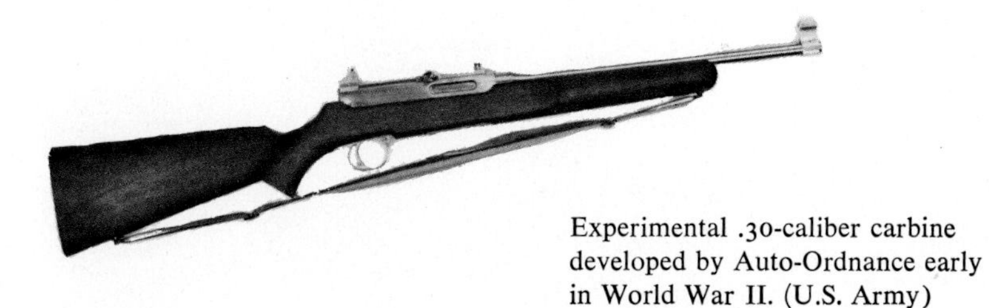

Experimental .30-caliber carbine developed by Auto-Ordnance early in World War II. (U.S. Army)

Pfc. William F. Doty firing a Thompson in a forest near the Siegfried Line in Germany in World War II. (U.S. Army)

American soldier reloading his Thompson under sniper-fire during the liberation of Paris. (U.S. Army)

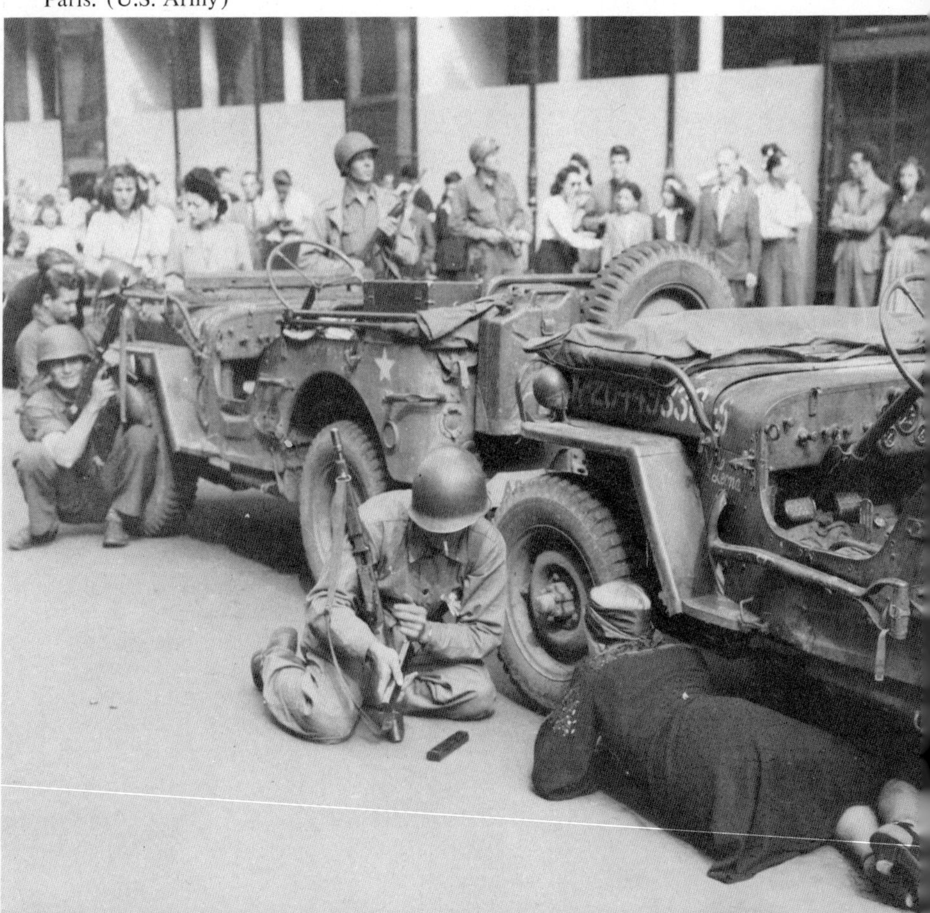

Advertisement using popular symbols of the Roaring Twenties: "gangster" suits and Tommyguns. (Used by permission of Canada Dry Corp.)

Numrich Arms Corporation employees, armed with Thompsons, posing playfully with 1920s cars and clothes. (George Numrich)

Red Chinese poster showing Vietcong guerrilla, armed with a Thompson, attacking American soldiers in Vietnam.

wrote a personal congratulatory letter to Maguire remarking on how much the British liked their Tommyguns, and the Republic of China purchasing agent wrote to express his country's gratitude: "Deliveries on your guns were ahead of schedule and once delivered, the guns themselves are probably the most valuable of all the ordnance pieces the Republic of China has received so far in any quantity. There was a time, you know, when your guns were about the only ordnance China was receiving. I rather think the Tommy gun has meant the difference between further retreat of the Chinese Army and their signal advance within the past few months."[4]

Auto-Ordnance achieved its high production rate by developing new manufacturing techniques, by operating its plants seven days a week with two eleven-hour shifts per day, and by bringing out a simplified model of the Tommygun that bore little more than a superficial resemblance to the weapon invented by John Thompson. The original Thompson for years had been obsolete in terms of contemporary weapons design. It was complicated, relatively heavy, and costly to produce both in time and money. It belonged to the First World War era that had inspired its development, and served in the Second World War only because Russell Maguire made it available in vast quantities that overshadowed all other considerations at the time.

When the Army at last saw a need for submachine weapons, it adopted the Thompson, but it also recognized the Thompson's obsolescence and began looking for a replacement. This did not greatly disturb Auto-Ordnance. The M1928A1 was the only submachine gun in mass production in any Allied country, and this gave the Thompson an enormous advantage over any would-be competitors. Starting in October 1939, the Ordnance Department tested several new submachine guns, and by the end of 1940 had found several that surpassed the Thompson in one way or another. But none improved on it enough to warrant adoption or justify the early production problems this would entail.

During 1941, however, the demand for Tommyguns increased so greatly that Auto-Ordnance felt compelled to simplify certain features of the gun in order to speed up production. The Thompson's elaborate Lyman rear sight consisted of ten separate parts and required about forty-five minutes to manufacture. The sight provided for both range and windage adjustments more appropriate to a long-range rifle than a submachine gun, and was replaced in December 1941 by a simple one-piece stamped battle sight that required only two minutes of machine time. Auto-Ordnance likewise discontinued the fancy checkering on the safety lever, the selective-fire switch, and the actuator knob. The barrel fins, smoothly finished on the original Colt-made guns, were left square-cut on the M1928A1, and were later dispensed with altogether. At the Army's insistence, Auto-Ordnance sacrificed the attractive lines of the walnut buttstock by reinforcing it with an unsightly bolt and washers.

But these modifications were minor compared to the changes proposed by the Savage Corporation. Engineers at Savage wanted to completely scrap the Blish lock, the very heart of the Tommygun, which they considered a frill if not a fraud. It added greatly to the cost and time of manufacture, detracted from the gun's reliability, and simply wasn't needed with the low-powered .45 ACP cartridge. Practically none of the dozens of submachine guns developed since the 1930s used any kind of breech lock whatever, relying entirely on the inertia of the bolt to seal the breech closed at the moment of firing. To prove their point, Savage engineers put together a straight blow-back Thompson and showed it to Auto-Ordnance.

Maguire, however, balked at such a complete remodeling of the gun. The Thompson's most publicized feature throughout its history had been its unique Blish breech-locking system, and Auto-Ordnance was not eager to admit that the Blish lock was superfluous. Which in fact it was, just as some Auto-Ordnance engineers had suspected when they were designing the gun in 1917 and 1918. While he was working on the temperamental Autorifle, Theodore Eickhoff had come to the conclusion that Blish's principle of metallic adhesion was a "complete farce," and he was amazed that the Patent Office had accepted the claims made for a breech lock that in his opinion worked (when it worked at all) on a principle no more arcane than simple friction magnified by mechanical disadvantage. Once, Eickhoff had tried to suggest this to John Thompson, as he felt was his duty. But to the old general, the Blish principle was too much an article of faith; he sent his young engineer back to the drawing board with a scolding that such talk could only undermine the very foundation of the company. Eickhoff never raised the topic again, taking some comfort in the fact that the Blish lock did absorb a small portion of the recoil and thereby gave the gun a smoother action.[5]

Russell Maguire's Auto-Ordnance wanted to retain the Blish lock because it was a traditional selling point, but Savage persisted, threatening to submit its own, simplified version of the Thompson to the Ordnance Department if Auto-Ordnance would not. Auto-Ordnance relented and produced a remodeled Thompson for testing around the end of 1941.[6]

The new model outwardly resembled the original Tommygun and incorporated the M1928A1's later modifications. But it completely eliminated the Blish lock in favor of straight blow-back, which in turn eliminated the separate actuator. This permitted the cocking knob to be mounted directly to the bolt and to project from the right-hand side of the receiver instead of from the top. In addition the gun had a permanently affixed buttstock of slightly different shape, a slimmer receiver, an unfinned barrel, no provision for drum magazines, and no Cutts compensator, which the Army had eventually decided was more glamorous than effective.

The new Thompson was tested in the early part of 1942. It proved superior to the M1928A1, inferior in some ways to other submachine guns under consideration, but again the most desirable weapon from a production standpoint. Its manufacture required no major factory changes, and the Ordnance Department standardized it in April 1942 as Submachine Gun, Caliber .45, M1.[7]

Soon the M1 Thompson was simplified even further by replacing the hammer system with a fixed firing pin machined onto the face of the bolt. Adopted in October 1942 as the M1A1,[8] the gun was now the simplest (if still not the cheapest) of "slam-fire" weapons. The bolt simply picked up, chambered, and fired a cartridge each time it closed, and was then blown back to repeat the cycle. The only moving parts were the bolt and the trigger-sear mechanism that either held or released it.

Modernizing the Thompson included scrapping its traditional magazines. The hundred-round drum, which weighed eight and a half pounds fully loaded, had been obsolete since the late 1920s, and the fifty-round drum, although issued at the start of the war, soon proved too bulky to carry and too hard to load under combat conditions. This left only the standard Type XX box magazine, whose twenty-round capacity was deemed inadequate for military purposes. In 1940 the Seymour Products Company, an Auto-Ordnance subcontractor, reportedly manufactured 25,000 box magazines of twenty-five round capacity, but these all seem to have vanished along with any records of their purpose or purchaser.[9] In 1941 the Ordnance Department studied magazines of both drum and box type. It also experimented with a special forty-round magazine consisting of two twenty-round magazines spot-welded together with top ends opposite and staggered sufficiently to permit insertion in the gun either way. This was an idea picked up from Allied combat troops who had already learned to improvise the same thing with electrical tape. This dual magazine arrangement was awkward to package and to carry, however, and in November 1941 the Ordnance Department settled for a lengthened version of the original box magazine holding thirty rounds.[10] The M1928A1 would take all types of Thompson magazines, but the M1 and M1A1 took box magazines only and were supplied with the thirty-round type.

The remodeling of the M1928A1 into the M1A1 cut the Thompson's production time by half and reduced its cost by 40 per cent. In 1939 Thompsons cost the government $209 each. By the spring of 1942, when the M1928A1 was on its way out, this price had been reduced to about $70. The M1 came into production at the Bridgeport plant in July 1942, and the M1A1 at the Savage plant the following December. By the time the last M1A1 submachine guns were manufactured for the Army, in February 1944, the unit cost of the Thompson had reached a low of $45, including accessories and spare parts.[11]

The "grease gun" replaced the Tommygun in 1944, but the Thompson had been doomed anyway from the very start of the war. It would not have fought World War II at all had Allied military planners anticipated the kind of war the enemy would wage and the important role submachine-type weapons would play in it. The war was one of paratroops, patrols, commando raids, motorized infantry, street fighting from house to house, and jungle combat. Short ranges and high firepower. Bullets sprayed into thickets or into rooms. The Germans had the light, inexpensive, efficient Schmeisser machine pistol in great quantity from the start. Allied countries made do with the heavy, costly, but dependable Thompson until each could develop and mass produce a weapon equally as good.

After the disaster (and miracle) at Dunkirk, the British ordered thousands of Thompsons and began designing a cheap submachine gun patterned on the German Schmeisser. The result was the STEN gun (for its inventors, Sheppherd and Turpin, and Enfield Armory), a crude but efficient "ten-dollar" Tommygun that went into production in the summer of 1941 and eventually reached a production rate of almost 47,000 a week.[12] Like the Schmeisser, it was 9mm and made largely of stamped parts. During 1941 and 1942 the Australians developed their own version of the STEN, called the AUSTEN, patterned even more closely on the Schmeisser. In 1934 the Soviet Union adopted the PPD submachine gun, which was manufactured in several models until superseded by the PPSh and PPS series after 1941. Unlike the Americans, British, and French, the Russians early recognized the submachine gun's value as an infantry weapon and used it as extensively as the rifle, but they needed thousands of Thompsons to supplement their own production during the first part of the war. The French army adopted the MAS (Manufacture d'Armes de Saint-Etienne) submachine gun in 1938, but never got it into mass production. France fell before it could take delivery on the six thousand Thompsons ordered from Auto-Ordnance during the winter of 1939–40.[13]

In October 1939 the United States Army opened a testing program in hopes of finding a cheaper, lighter submachine gun to replace the Thompson.[14] Because it already had the Thompson and did not yet have a war, the testing was carried out at a fairly leisurely pace and confined to guns under development by private firms or inventors. The first of these was the Hyde Model 35, an improved version of the Model 33 which the Lake Erie Chemical Company had tried to market in the early Thirties in hopes of capturing some of Federal Laboratories' strike-breaking and revolution business. Its inventor was George Hyde, a German machine-gun designer during World War I who emigrated to the United States in 1926 and later worked as a gunsmith for the New York sporting goods firm of Griffin & Howe. At his Brooklyn home in his spare time, Hyde worked on his submachine gun and eventually built about sixty guns for the Lake Erie Company.[15] Both models of the gun closely resembled the Thompson in outward appearance, and the Model 35 surpassed the Thompson in several respects, but it was found to have flaws in both design and construction. Later in the testing program the Atmed Submachine Gun Company of New York submitted another gun designed by Hyde, but like the Model 35, it was not superior enough to the Thompson to warrant adoption.

Other guns tested by the Ordnance Department were the Spanish Star, the Finnish Suomi, the Reising (Harrington & Richardson Arms Company), the High Standard (Gus Swibelius, High Standard Manufacturing Company), a semiautomatic carbine developed by Smith & Wesson, and the British Mark II STEN.

The Army's first series of tests ended in January 1942. Although none of the entries was adopted, some of them impressed the Ordnance Department favorably and proved that a better submachine gun than the Thompson could be designed. On a scale of 100, measuring reliability, simplicity, accuracy, and weight, the M1928A1 Thompson scored only 57, while the Hyde scored 70, the Hyde-designed Atmed 77, and the High Standard 73. To the Army's surprise the highest score of all, 88, was earned by the British Mark II STEN.

When first presented, the STEN offended American military sensibilities. Stamped out, welded together, painted instead of blued or Parkerized, it was embarrassingly crude and graceless. In competitive tests, however, it surpassed the Thompson in most respects, and could be manufactured in a fraction of the time and at a fraction of the cost. While slowly absorbing this revelation, the Ordnance Department embarked on a new series of submachine gun tests which this time would include not only privately invented weapons, but also designs developed under Ordnance Department contract.

This was a departure from Army policy that had deterred the Ordnance Department from trying to develop a submachine gun on its own. Government armories were neither staffed nor equipped to develop new weapons quickly and from scratch, and even if the Department had done so it still would have had to farm out the manufacture to private industry to meet wartime production needs. In addition, the Ordnance Department would have had to risk large expenditures of time and money on a new design that might well flop, and then start all over again. Everything considered, it had seemed wisest to turn the entire matter over to private industry—to stage a big free-for-all competition, and then pick the choicest fruit of everyone's combined time and talent.[16] The incentive to enter such a contest was the prospect of enormous wartime royalty payments, since inventors retained patent rights to adopted guns. It was this incentive that induced John Thompson to undertake the private development of a semiautomatic rifle in 1916, and attracted Russell Maguire to Auto-Ordnance in 1939.

But this competitive system also had its drawbacks. In the first place, the money and man hours spent by all the losing contestants was largely wasted. Secondly, by setting itself up as an impartial judge, the Ordnance Department had to refrain from prompting any one contestant with suggestions for improving his design, other than pointing out its defects. Finally, the public advertising of military design requirements kept the enemy informed on what to expect in the way of new weapons.[17]

In view of these considerations, and also of the failure of private inventors to submit any outstanding designs, the Ordnance Department in late 1941 decided to take a more active part in the developing of a new submachine gun. It did not, however, turn to its own armories, which were busy enough already. Instead the Department contracted with the Inland Manufacturing Company of Dayton, Ohio, and with George Hyde to combine their efforts toward a submachine gun incorporating the best features of the weapons then being tested. Hyde's earlier guns were sound and simple in design, and Inland, which was producing Army carbines, had the small-arms manufacturing experience to solve construction problems.

The new Hyde-Inland submachine gun was completed in prototype in April 1942, and submitted to the Ordnance Department. It did not resort to the extensive use of metal stampings, as did the STEN, and its full rifle-type stock gave it a conventional appearance. But it eliminated a good deal of machining through the use of seamless steel tubing for the receiver and some internal parts made by powder metallurgy, and it scored a high 84 on the Ordnance Department's testing scale. On the basis of prototypes the Hyde-Inland was adopted in April 1942 as substitute standard (to supplement the M1 Thompson) and designated U.S. Submachine Gun, Caliber .45, M2.[18]

In addition to being cheaper and simpler, the M2 had one conspicuous advantage over the Thompson and most other submachine guns of the day. Owing to the design of its buttstock it had almost no tendency to climb off target during full-automatic fire. In tests the Ordnance Department had discovered that muzzle rise was the result of a gun's recoil force acting along a line higher than its point of support. Since most guns were supported by a buttstock angling downward to a point well below the boreline, they had a tendency to pivot upward with each shot. The solution was simple enough: raise the buttstock until it was in line with the bore so that the force of recoil acted in a straight line against the shoulder. This was demonstrated by means of a modified Thompson M1928A1 whose stock extended straight back from the receiver.[19] The experimental Thompson, however, required greatly elevated sights that would not have been practical for military purposes. The M2 sacrificed perfect alignment of barrel and buttstock to avoid such a fragile sight arrangement, yet managed to raise its stock high enough to nearly eliminate muzzle climb.

On paper and in prototype the M2 seemed measurably superior to the Thompson, and in July 1942 the Ordnance Department contracted with the Marlin Firearms Company of New Haven to rush the gun into production. Marlin had both the experience and the plant capacity to manufacture submachine guns. The company had just finished building fifteen thousand United Defense submachine guns (as the High Standard gun was called) for the Dutch. But when Marlin tried to mass produce the new M2, so many design and materials problems arose that the program bogged down completely. The Army had to go back to Auto-Ordnance for more Tommyguns.

While Marlin Firearms struggled to get the contrary M2 into production, the Ordnance Department continued testing new submachine guns submitted by American designers. It also took a long second look at the British STEN (Mark III), and at the Australian AUSTEN and the German Schmeisser. When none of the American designs, including the M2, could match the cheap, homely foreign weapons for utter practicality, the Ordnance Department conceded in a test report that "in modern warfare there are other criteria than mere appearance."[20] Whereupon the Department sent George Hyde back to his drawing board at Inland to develop an all-metal submachine gun just as cheap and effective as the British STEN, and if necessary, just as ugly.

Development of the new gun was authorized by an Ordnance Department directive in October 1942. A month later Inland furnished the Army with a working prototype that met all requirements practically to the letter. It was smaller than the Thompson, weighed about two pounds less, and could be manufactured for a little under $20 in half the time and with a 25 per cent reduction in machine-tool requirements. It was more reliable than the STEN under adverse conditions, as accurate as the M2, and fired at a desirably slow rate of about four hundred shots per minute. On December 24, 1942, after only a month of testing, the new gun was adopted as standard and designated U.S. Submachine Gun, Caliber .45, M3.[21] The Thompson and the M2 were both downgraded to limited standard, and six months later the M2 program was scrapped altogether.

The Army had adopted the M1 Garand semiautomatic rifle in 1936 after years of leisurely testing and refinement. Under wartime pressure, with the benefit of preliminary research and foreign examples, the Ordnance Department rushed the M3 from design to adoption in less than two months, and into production four months later. In both appearance and construction the M3 bore an uncanny resemblance to the ordinary grease gun and it soon acquired that nickname. The old Thompson remained the Cadillac of submachine guns; the M3 was the Jeep.

The M3 was put into production in the summer of 1943 by the General Motors Corporation's Guide Lamp Division, whose machinery for making automobile headlights was well suited for stamping out the new submachine gun. During the last two years of the war about 600,000 were manufactured for the Army. But while the M3 proved itself to be the ideal submachine gun for military purposes, it ran into the usual manufacturing difficulties at first and was six months in reaching full-scale production. Again the Army had to turn to the Auto-Ordnance Corporation for more Tommyguns, renewing canceled orders for 120,000. Auto-Ordnance had been notified in January 1943 that the Thompson was superseded by the M3, but it was thirteen months later when Tommygun production actually ceased. In February 1944 the Army took delivery on its last 2,091 Thompsons, to bring its procurement since 1940 to a total of 1,387,134 in models M1928A1 (562,511), M1 (285,480), and M1A1 (539,143).[22]

The M2 submachine gun surpassed the Thompson in theory and in tests, but could not get far enough off the drawing board to replace it. The M3 eventually supplanted the Thompson, but it did not reach American troops in large numbers until 1944. Due to a scarcity of Thompsons in 1942 the Marine Corps adopted the Reising, made by Harrington & Richardson, but these guns proved so unreliable that Marines on Guadalcanal reportedly threw them in the Lunga River and made do with rifles until they could get more Tommyguns.[23] By ordnance standards of World War II the Thompson submachine gun was an antique. But because it was such a good one, and for a long time the only one available to the Allied armies, it became one of the most widely used small arms of the war and certainly one of the most famous.

Most of the Thompsons maufactured from 1939 through 1941, including those ordered by the Army, went to Britain and other Allied countries. Such military aid had been prohibited under the terms of the neutrality acts of the 1930s. The acts clamped an automatic embargo on arms to any warring countries and resulted in the cancellation of $79,000,000 worth of French and British military orders at the outbreak of the war. But Germany's early and impressive victories frightened the United States out of its total isolationism, and in November 1939, Roosevelt managed to secure a revision of the neutrality laws which permitted the sale of war matériel to belligerent countries on a strictly "cash and carry" basis. Since the British still ruled the sea, at least on the surface, this revision constituted a loophole through which only Allied shipping could pass. The United States government even expedited matters by purchasing arms, declaring them surplus, and then turning them over to private firms for resale to Allied purchasing commissions. After 1940, when the exhaustion of British dollar reserves ruled out any more "cash and carry," Roosevelt evaded the ill-conceived neutrality law altogether with his famous Lend-Lease plan, which became law in March 1941. Lend-Lease permitted the Allies to pay for munitions not in dollars but in goods and services at the end of the war, and through Lend-Lease thousands of Thompson submachine guns went to virtually every Allied army and into every theater of combat. The Chinese, the British in North Africa, the Australians in the South Pacific, and the Russians at Stalingrad depended heavily on the Thompson and held it in high regard as a weapon.[24]

The British commandos probably did the most to glamorize the Thompson, adopting it as part of their insignia and using it in their daring night raids against German installations on the coasts of Norway, France, and North Africa. News pictures of the fighting in North Africa almost inevitably showed British troops and tank crews carrying the commercial Model 1928 Thompson with its familiar drum magazine and pistol foregrip. In the desert campaigns, in the defense of Crete, in the storming of the French naval base on Madagascar, the British used the Thompson so widely and so conspicuously that many Americans still think of the gun as primarily a British weapon, nicknamed for the British "Tommy."

In July 1940 the United States Army had on hand only 260

Thompsons,[25] and it probably didn't have many more than that when the Japanese attack on Pearl Harbor drew the country into the war. Almost every gun manufactured had been sent to Britain in response to cable requests for 25,000 submachine guns "urgently required to meet parachute attacks expected in the early future."[26] After Dunkirk, the British wanted all the Thompsons they could get, as did the Russians. After the Pearl Harbor attack, the Ordnance Department rushed thousands of Thompsons to the Army and the Marines in the Pacific. Despite the increasing availability of the M3 after 1944, the Thompson stayed in service both in Europe and the Pacific until the end of the war.

The Ordnance Department supplied 183,973 Thompsons to the Navy,[27] and probably that many or more to the Marines. The Navy's guns were used mainly by landing parties and by the Construction Battalion, the famous "Seabees," whose emblem pictured a fierce-looking bee carrying a drum-magazine Thompson. The emblem of the Army and Navy Amphibious Forces also included a Thompson, as did numerous magazine covers, posters, and advertisements. Nor was any war movie complete without a few spectacular Tommygun scenes.

In most of the war's important land battles, as well as the exploits of individuals, the Thompson figured prominently. General Frank Merrill's "Marauders" used Tommyguns in their commando-style raids behind Japanese lines, and Lieutenant General Joseph Stilwell personally carried a Thompson through Burma. In 1944 the Navy's first official "boarding party" since the War of 1812 used Tommyguns in capturing the German submarine U-505 off the coast of West Africa.

During the war the Thompson gradually lost its gangster image as the familiar pistol foregrip and drum magazine gave way to the straight forearm and long thirty-round "clip." It looked even less like a "chopper" when the M1 replaced the old "Al Capone" model. But its former reputation never disappeared altogether. A Russian Army submachine gun manual referred to the Thompson as a "Chicago typewriter," and a British manual identified it as the "original American gangster-gun." In 1942 the British edition of *Yank* published an anecdote about an Air Corps sergeant who seemed unusually familiar with the Thompsons that had just been issued to his squadron. He quickly stripped one down and reassembled it, then nonchalantly put it to his hip and shot a target to pieces. In reply to a major's question, he explained, "Well, sir, you see, I once hadda take these things apart in the back of a car going seventy miles an hour." The sergeant turned out to be a Chicagoan, according to the punchline, who had once served five and half years for bootlegging.[28]

Even when the M3 and the STEN began to replace the Thompson in the last year of the war, "Tommygun" remained the popular term for any kind of Allied submachine gun. In reporting the battle for Berlin in April 1945, the Associated Press described Russian "Tommy gunners, red as Indians from clouds of brick dust rising from the debris," and war correspondent John Groth described the "sharp crack of rifles, the staccato burping of Tommy guns."[29] The Red Army was equipped mainly with Russian-made weapons by the end of the war, but in American journalistic accounts of battles any kind of submachine gun was still a "Tommy."

The Thompson's merits as a military small arm, especially compared to other submachine guns, remains a question over which gun buffs and ex-GIs still argue. War stories both praise and condemn the gun: American paratroops, as soon as they could get their hands on a German Schmeisser, broke their Thompsons over the nearest rock; the roadside from Salerno to Naples was littered with Thompsons discarded by soldiers tired of carrying the heavy things; the Thompson was such a desirable weapon that infantry soldiers often bought them or stole them from tank crews; when the new M3 arrived, many Marines would not turn in their battered but faithful Thompsons. One widely told and widely believed story is that British troops in North Africa (or American soldiers elsewhere) learned to increase the reliability of the M1928A1 by taking out the Blish lock and throwing it away. The point of the story usually is to illustrate that the famous Blish breech-locking system only made the gun more complicated and therefore more susceptible to jamming from sand, mud, or whatever. The Blish lock was indeed superfluous, as proven by the Thompson M1 models, but it was never simply removable. It connects the actuator to the bolt for the purpose of cocking the gun.

As a combat weapon, the Thompson had its conspicuous shortcomings—chiefly its size, weight, and cost compared to the all-metal submachine guns made largely from stampings. But most of the other complaints against it could be written off, as the Army discovered, to the normal soldier's normal tendency to consider his own weapons inferior to those being fired *at* him. Similarly, exaggerated fondness for the Tommygun usually sprang from a soldier's sense of the dramatic, or from a sentimental possessiveness toward that which was *his* and had helped him get out of a few tight spots. Throughout its military history, however, the Thompson has been good to a fault in one respect: it draws fire from enemy soldiers or guerrillas eager to get their hands on one. Like any weapon, the Thompson's merits depended a lot on the man using it. Realistically, it was a sound old gun that was available when desperately needed.

10

Antique and Obsolete

> I once sat, a prisoner, long ago, and watched a peasant soldier just recently equipped with a submachine gun swing the gun slowly into line with my body. It was a beautiful weapon and his finger toyed hesitantly with the trigger. Suddenly to possess all that power and then to be forbidden to use it must have been almost too much for the man to contain. . . .
>
> "Thompson, Tome′-son′," he repeated proudly, slapping the barrel. "Tome′-son′." I nodded a little weakly, relaxing with a sigh. After all, we were men together and understood this great subject of destruction. And was I not a citizen of the country that had produced this wonderful mechanism?
>
> —*Loren Eiseley, The Immense Journey*

After two decades of notoriety and financial misfortune, the Tommygun distinguished itself during the Second World War and then passed quickly into history. In some respects it did not even survive the war. The Army's M1 models, though still resembling the original, were Thompsons only in name and a few interchangeable parts, and with their introduction in 1942 the new Auto-Ordnance Corporation, Russell Maguire's, severed its last connection with General John T. Thompson and the past. The Thompson Automatic Arms Corporation, organized for the purpose of purchasing Auto-Ordnance from the Ryans, was absorbed by its own subsidiary in October 1941. In March 1944 Auto-Ordnance itself became a subsidiary of a new corporation, Maguire Industries.

Under Maguire's capable management the company thrived. By the end of 1943 Auto-Ordnance plant area totaled 301,680 square feet of floor space, and the company's employees numbered about three thousand. Gross income increased from $171,511 in 1939 to a one-year high of $42,610,293 in 1942. The Thompsons fast receded into obscurity as Maguire became the country's "Tommygun tycoon." By 1944 Auto-Ordnance had sold the United States and the Allies almost $130,000,000 worth of submachine guns and parts, earning the company a net profit of some $14,845,000, and earning Maguire personally about $2,000,000 on his original investment of $100,000.[1]

Maguire's road to success was not without a few rough spots. Soon after he acquired control of Auto-Ordnance, Thomas Kane and Evelyn Thompson filed suits to recover the stock they felt Maguire had extorted from them with his zero-hour ultimatums. Maguire's attorneys worked diligently to keep the case from coming to trial, and on September 16, 1940, Maguire declared a $5.50 dividend on Auto-Ordnance stock, payable the same day. This caused a commotion on Wall Street, and also upset Kane and Evelyn Thompson. The usual practice was to announce a dividend payable ten days later, which allowed time for processing any pending stock transfers. Kane rushed to court and obtained an injunction against payment of the dividend, which he argued would permit Maguire to strip the corporation of more than $600,000 on stock that he stood a good chance of losing when the Kane-Thompson suit finally got to trial. Rather than contest the injunction in court and give Kane a forum, Maguire consented to the entry of the restraining order. In February 1941, after much bargaining, both sides reached an out-of-court settlement that amounted to partial restitution of the stock acquired from Kane and a cash payment to the Thompson estate.[2]

In 1942 Maguire was sued by Richard Cutts, Jr., and Lawrence

E. de S. Hoover over royalty payments on the Cutts compensator. Hoover had been employed as an Auto-Ordnance sales representative from 1923 to 1926. During that period he had met Cutts and his father. When Maguire gained control of the company in 1939, Hoover, who had good business contacts in military circles, rejoined the company as its Washington sales representative; he also made a private deal with Cutts to act as exclusive agent in the sale of compensators.

According to agreements signed in 1940, Auto-Ordnance was to pay Cutts $1 royalty per compensator, up to a maximum of $250,000. However, when Tommygun sales zoomed during 1941 and 1942—650,000 compensators worth—Cutts realized he had sold too low. Hoover likewise wanted a bigger share of Auto-Ordnance profits. He had taken the job in 1939 at $8,000 a year; he would stay on the job after 1941, he informed Maguire, for $100,000 a year, plus 51 per cent of any bonuses paid to the top three officials of Auto-Ordnance. Now Maguire found himself issued an ultimatum: accept Hoover's demands within forty-eight hours or lose the only man who had any knowledge of the company's affairs as they related to the government and the military in Washington. Hoover also hinted that he could give the War Department certain information that would be most damaging to Auto-Ordnance.

Maguire apparently agreed to Hoover's demands but never met them, and furthermore tried to deal directly with Washington despite Hoover's exclusive agency. Charging that Maguire had misrepresented certain facts (presumably sales prospects) in making the original agreements, and that Auto-Ordnance had breached a collateral obligation with regard to Hoover's agency rights, Cutts contended that he therefore was not bound by the $250,000 maximum already paid him. The case was submitted to arbitration, Cutts receiving $417,790 in damages and Hoover $30,000 on his own claims. Maguire contested this settlement in court, however, and succeeded in getting the awards reduced to a total of $250,000.[3]

The information Maguire feared Hoover would spill to the War Department probably concerned pricing, for in 1942 Auto-Ordnance found itself in difficulties with the government over excessive profits, thanks partly to Hoover and Cutts. As an officer in Auto-Ordnance, Hoover had inside knowledge of costs and prices and considered them far out of line. By 1941 mass production techniques had greatly cut the Tommygun's cost of manufacture, but these savings had not been passed along to the government. Cutts believed that Auto-Ordnance was getting far too high a price for compensators. When they could not persuade Maguire to adjust prices, Hoover and Cutts brought the matter to the attention of the Ordnance Deparment, only to discover that Auto-Ordnance prices were already being scrutinized.[4]

As a result of the Army's investigation, Auto-Ordnance contracts were renegotiated in 1942 to bring the government not only a substantial price reduction but also a refund of some $6,500,000. On the basis of final audited figures for the year, Auto-Ordnance was assessed an additional $750,000.

These assessments were by no means unusual. During that same year many government contracts were renegotiated, some for much greater amounts. Viewing himself more as a defense production hero than a war profiteer, Maguire applied in the tax courts for redetermination of the refunds, and also sued the government in federal claims court to recover $8,000,000 in refunds and tax credits. He lost both cases, however, and the government won an excessive-profits judgment of $258,800 against Auto-Ordnance in 1943.[5]

Despite his great success with the Tommygun, and even before that success was assured, Maguire had intended to market other military small arms. This was initially a hedge against the chance the Thompson might not hit the jackpot; later, against the day the Thompson inevitably would be superseded by some better weapon.

Late in 1940 the Auto-Ordnance office in New York received a visit from William B. Ruger, a young inventor barely out of his teens who would one day be well known for his excellent handgun designs. Ruger had built a crude model of a light machine gun, a class of weapon falling somewhere between the air-cooled Browning medium machine gun and the Browning automatic rifle, and last represented by the old Lewis gun. Since the War Department had sent a memorandum to arms designers in hopes of inspiring just such a weapon, Auto-Ordnance and Ruger quickly got together on a contract. However, young Ruger and young Auto-Ordnance were about equally inexperienced in the business of developing new machine guns. After several months, working eighty-hour weeks sequestered in his private corner of the new Bridgeport plant, Ruger discovered that his finished, working model did not work well at all. It received enough compliments in government tests that Auto-Ordnance staked Ruger to another two years' work on the gun, but as the war went on, the Army lost much of its earlier interest in such a weapon, and so did Ruger and Auto-Ordnance.[6]

While Ruger worked on his light machine gun, Auto-Ordnance was also sponsoring efforts to develop a .30-caliber carbine rifle. An elderly designer named Grant Hammond labored in another corner of the Bridgeport plant to convert an automatic pistol, which he had developed years previously, into a short-recoil carbine that would meet the specifications listed in another of the Army's small-arms memorandums. The Hammond carbine reportedly showed considerable promise, but eventually lost out in competition with the Winchester Garand-type entry, which the Army adopted as the M1 carbine.[7]

In late 1941, although Thompson orders were piling up into the hundreds of thousands, Auto-Ordnance began taking steps to develop a new submachine gun before some competitor came up with an improved design the Army would find irresistible. First, Maguire made a token effort to promote the Model 1929 B.S.A.-designed Thompson, both in its original form and in a modified version refurbished with unfinned barrel, Bridgeport markings, and no compensator. The latter, chambered for 9mm, was apparently aimed at the Lend-Lease market. But since both the B.S.A. and the Bridgeport prototypes retained the Blish breech system, neither aroused much interest. When Savage and the War Department compelled Auto-Ordnance to discard the Blish lock, the company came up with still another 9mm prototype, which combined certain features of the old M1928A1 and the simplified M1 models. However, it could not compete with the cheap, stamped-metal designs already coming into fashion.[8]

Quite aware that the Army had approved the M1 Thompson only

for lack of any cheaper, simpler gun, Auto-Ordnance stepped up its efforts to come up with something altogether new. Another Hammond in the engineering department, William Douglas Hammond of Los Angeles, designed a straight blow-back tubular receiver submachine gun closely resembling the Hyde-Inland M2, then being tested by the Ordnance Department. Auto-Ordnance designated the Hammond gun the Model T2 and submitted it to the Army in October 1942. The Auto-Ordnance Model T2 excelled in semiautomatic accuracy, but scored lower than the M2 and no better than the M1A1 Thompson in other parts of the test.[9]

When the Army, after adopting the M2, changed its mind and issued a call for an all-metal submachine gun, Auto-Ordnance put W. D. Hammond back to work on a weapon along those lines. Over the next few months he designed several such guns, some employing what was then a very novel construction principle: the entire receiver and grip structure made from two heavy-gauge drawn halves meeting at a vertical centerline, a method common in toy manufacture.[10] However, the Hammond all-metal submachine gun was not completed in time to compete with the Hyde-Inland M3, which the Army adopted as its long-sought successor to the Thompson.

Auto-Ordnance never succeeded in marketing any of its World War II designs (other than the M1 and M1A1 Thompsons), but it did engage in subcontract work for the government on other small arms. During the latter part of the war, the company performed some of the assembly work on the M1 carbine, and also manufactured barrels and other components for the Browning automatic rifle.

Due to wartime pressures, Auto-Ordnance made few efforts to modify or improve the Thompson in ways not related to speeding up production. The company's experimental guns, and Thompsons built experimentally in other calibers, were motivated solely by existing military needs. During the early part of the war, however, a private inventor made one last stab at an old idea—converting the Thompson to fire .22-caliber rimfire ammunition. In 1942 Charles William Robbins, owner of a small shop and ordnance laboratory in Cincinnati, filed for a patent on a conversion unit which not only adapted the Tommygun to .22, but retained most of the gun's recoil and "feel." The Robbins unit consisted of five main parts: .22-caliber barrel insert, bolt, lock-frame, buffer, and magazine. These parts were merely substituted for the gun's standard parts, without tools or alterations, and could be replaced with the standard parts in about a minute. The purpose of the Robbins adaptor was to provide a means of using inexpensive .22 long-rifle ammunition for police and military submachine gun training. Wartime priorities apparently ruled out any serious military interest in the device, however, and only a few were produced for demonstration purposes.[11]

Knowing almost at the start that the Tommygun's days were numbered, Maguire anticipated the decline of his ordnance business in time to diversify his company. As early as 1941 Auto-Ordnance acquired an interest in several Kansas oil wells. In 1943 the company expanded its oil operations into Texas, and also branched out into the manufacture of electrical equipment and compressed food products. In March 1944, a month after the last submachine gun came off the Bridgeport assembly line, Maguire reorganized the company. The Auto-Ordnance Corporation was downgraded to become merely the Ordnance Division of a new parent company, Maguire Industries, Incorporated. At the end of the war the Bridgeport factory was stripped of its gunmaking machinery and retooled for the manufacture of radios, record players, and other scarce consumer items.[12]

Despite its preparations for peacetime business conditions, Maguire Industries fared poorly during the immediate postwar years. The compressed-foods market dried up, and Maguire's electrical components business had difficulty adjusting to the loss of its defense contracts. Also, the company still owed money to the government.

Maguire himself ran into trouble with the SEC when he tried to go back into the securities business in 1945. Before the war the commission had forced him to dissolve two securities companies when it caught him manipulating the stock of the Sterling Aluminum Products Company of St. Louis. Maguire, his wife, two clerks, and an attorney were charged with making "wash sales on the curb exchange" (as one newspaper[13] phrased it) while pushing the stock at lower prices "over the counter." When Maguire tried to register a new securities company, Maguire, Incorporated, in 1945, the SEC turned down his application with a stern scolding:

> The prior revocation proceedings were based on flagrant violations of the anti-manipulation and anti-fraud provisions of the laws we administer. They were neither casual nor inadvertent. They were planned, willful and persistent.
>
> They bespoke, not ignorance of the law, but a knowledge of its requirements and an attempt to use devices [dummy accounts] to evade it. Revocation would have been necessary in the public interest had not the respondent dissolved and assured us that Maguire had no intention of going back into the securities business. Maguire has not lived up to this assurance and has given us no reason for the change.[14]

Maguire Industries had to struggle through the late Forties. In the early Fifties the company started showing a profit, thanks partly to having branched into the lucrative manufacture of asphalt additives and emulsifiers for the paving industry. In 1961 the company was renamed the Components Corporation of America, with headquarters in Mount Carmel, Illinois. Its principal business is the manufacture of replacement parts for television and radio receivers, both commercial and military.

In 1951 Maguire opened a new, final, and somewhat dismal chapter to his already adventurous career by purchasing the *American Mercury* magazine. After its spectacular early years under founder-editor H. L. Mencken, the *Mercury* had withered in a Depression climate unfavorable to its 1920s-style wit and intellectual polemics. It was purchased in 1939 by Lawrence Spivak, producer of NBC's "Meet the Press," who sustained it another ten years as a literate, respected, but profitless journal of public affairs. In 1950 Spivak sold the *Mercury* to William Braford Huie, who turned it into a responsible conservative political journal and an even greater money-loser. Complained Huie, "The left-wing journals always seem to find someone's crazy grandson to finance them."[15]

Huie began looking for a crazy grandson of right-wing persuasion, and found Russell Maguire, who apparently had weathered the postwar years better than his company (*Newsweek* estimated him to be worth $100,000,000 in 1952). Maguire was a generous contributor to charities, and had been since World War II. "Money doesn't mean anything to me," he told one reporter. "I've given away hundreds of thousands."[16] When he agreed to bankroll Huie's *Mercury*, to open the nation's eyes to the threat of Communism, he explained, "I've done very well in America, and now I want to start putting something back."[17]

Maguire and Huie soon had a falling out over the *Mercury*'s editorial policies, as did Maguire and several later editors, who likewise felt he was trying to make the magazine an organ of the lunatic fringe. In 1955 most of the editorial staff resigned in protest over Maguire's extremist views and anti-Semitic editorials. Maguire quickly restaffed the magazine with bona fide crackpots and laid siege to the evils of the day: the United Nations, Jews, mental health programs, Dwight Eisenhower and other notorious Communist sympathizers.[18]

Maguire disposed of the *Mercury* about 1960, placing it in the capable hands of a Texas organization calling itself the Legion for the Survival of Freedom. Six years later, on November 10, 1966, Maguire died at his home in Greenwich, Connecticut, at the age of sixty-nine.

At the end of the war Maguire had no further use for his ordnance division. The special machinery, tools, blueprints, leftover parts, records, and everything else pertaining to the Tommygun went into crates and then into storage. For four years Auto-Ordnance remained a valueless paper corporation sealed in wooden boxes that only took up warehouse space.

In the early part of 1949 Maguire Industries received an unexpected offer to take the Tommygun off its hands. The offer came from the Kilgore Manufacturing Company of Westerville, Ohio, a firm nostalgically remembered by youngsters of the 1930s for its Kilgore cap pistols—made before caps and metal toys "went to war" and the company turned to military and commercial pyrotechnics. Kilgore had no interest in Thompson submachine guns as such, but had business contacts in Egypt that felt certain the equipment and drawings for making Thompsons could be resold to the Egyptian government at a handsome profit. Kilgore was confident of its resale market, and Maguire Industries was more than happy to dispose of Auto-Ordnance. The deal reportedly was opened and closed in only forty-eight hours, for the tidy sum of $385,000.[19]

Then Kilgore discovered that it had been a little hasty. According to one hearsay account, "The only thing Kilgore ever sent to Egypt was a couple of idiots"—two representatives who found the Egyptian government much less eager than Kilgore had believed, and who soon scuttled the deal completely by employing "used-car" salesmanship.[20] Having no use whatever for its newly acquired property, and no other prospective customers, Kilgore ruefully accepted the offer of a New York investment syndicate, headed by onetime Auto-Ordnance executive Frederic A. Willis, to take Auto-Ordnance off its hands at a cut-rate price.

If Kilgore regretted its one misadventure in arms speculation, the company soon had reason to wish it had never heard of Tommyguns. In October 1951 Frederic Willis sold Auto-Ordnance to the Numrich Arms Corporation of Mamaroneck, New York, a small firm specializing in surplus ordnance equipment and obsolete gun parts. Numrich wanted Auto-Ordnance to add to its growing collection of defunct arms companies (Standard, Forehand, Hopkins & Allen) whose assets usually consisted of the company's name and its stock of spare parts.

When George Numrich picked up the crated assets of Auto-Ordnance, he was surprised to discover that some of the boxes contained complete submachine guns—some eighty-six in all, including several of the very earliest prototypes. To comply with federal law, Numrich immediately registered the guns with the Treasury Department. This resulted in tax claims of $200 per gun against the Willis investment syndicate and against Kilgore Manufacturing Company, both of which had unknowingly sold full-automatic weapons without a Class Four federal "machine gun" license. Since neither Willis nor Kilgore had so much as looked at the stored crates, the government conceded that they had acted in ignorance and innocence. But the law was the law. The Willis group apparently was not penalized, but the Treasury nicked Kilgore for about $12,000.

Since buying Auto-Ordnance, Numrich Arms Corporation (now located in West Hurley, New York) has grown into the country's largest supplier of replacement parts for antique, obsolete, and military firearms. The company also manufactures modern muzzleloaders for gun hobbyists and operates a sizable retail store stocked with new and used guns of all types. In addition, Numrich still builds, sells, and services Thompsons. Numrich's "new" Thompsons are identical to earlier models and are made up largely from surplus parts, with the exception of frames and receivers. These are machined as needed in the company's shop and bear most of the original Auto-Ordnance markings. The initials "N.A.C." have been added to the serial number.

The Tommygun constitutes only a small, nostalgic part of the Numrich operation today. Occasional orders still come in from police and sheriffs' departments, and a few of the guns are still exported to foreign countries, mainly in the Near East. In the middle 1950s the Sheik of Kuwait purchased a relatively large batch of one hundred Thompsons and insisted that they all be the traditional 1921 and 1928 model "Tommyguns." The World War II M1 and M1A1 models did not have the right look about them. From time to time Numrich receives orders from private individuals willing to pay the $200 federal tax for the privilege of owning an operable Thompson submachine gun. Some of these buyers are serious collectors who simply want every piece in their collection to be shootable. Others are sportsman-romantics who want to machine-gun sharks in the Ernest Hemingway tradition, or strafe prairie dogs from low-flying airplanes. Texas millionaire sportsman and racing-car builder John Mecom reportedly tours his ranch in a half-track hunting bobcats with a Thompson.[21]

George Numrich has a romantic streak of his own, to which he attributes his interest in the Tommygun. From a strictly business point of view, Auto-Ordnance hardly earns its keep. But Numrich possesses a sense of history and likes to think of himself (at least jokingly) as a successor to Sir Basil Zaharoff, "Sidewalk" Sedgley, and the other colorfully unscrupulous characters from the munitions industry's past. His sense of humor runs along similar lines. One year recently the company's official Christmas card depicted Numrich and his employees at their assembly line, making guns out of plowshares. Numrich can afford the luxury of such humor because his conscientious business practices have kept him on good terms with the Treasury Department's Alcohol, Tobacco, and Firearms Division of the Internal Revenue Service, which enforces federal firearms laws.

By way of avoiding trouble, Numrich has never sold "deactivated" submachine guns—which have a way of getting reactivated and into difficulties of one kind or another. During the late Forties and early Fifties the Treasury Department inadvertently started this fad by campaigning to get dangerous souvenirs of World War II and Korea out of circulation, or at least rendered harmless. Persons owning prohibited weapons could keep them legally if they brought them in to be duly registered and the barrels welded closed (which included welding to the receiver to prevent their replacement). It soon became apparent that American GIs had been more resourceful at smuggling home guns than anyone had imagined; the Treasury found itself inundated with weapons and paperwork. To streamline things, the Treasury Department, in September 1955, instituted its Deactivated War Trophy (DEWAT) program. Once a gun was made inoperable under Treasury supervision, it was legally transformed from a firearm into a "trophy" of no further interest (or paperwork) to the government.[22]

It was not long, of course, before enterprising dealers in surplus arms discovered the DEWAT program to be an ideal means of dumping excess stocks of obsolete machine guns on the collectors' market, and the country soon was flooded with "dewats"—STEN guns, Schmeissers, Thompsons, Lewis guns, and many others—selling anywhere from $10 to $50 each. Nor was it long before curious gunbuffs, but also juvenile gangs and other undesirables, discovered that it was often possible to "rewat" such trophies. Some had been poorly welded in the first place. Because it broke the hearts of many a gunshop welder to "ruin" a perfectly good Thompson or Schmeisser, some took secret pride in their ability to do a convincing but superficial job. But even a gun that was properly welded usually could be restored to working order with enough machine shop facilities or simple determination.

By 1958 it was clear that the DEWAT program was a Frankenstein midget with considerable potential. Reactivated guns were beginning to turn up in crimes, in street-gang arsenals, and in the hands of foreign revolutionaries, including the indefatigable I.R.A. A *Saturday Evening Post* editor exposed this evil in the time-honored fashion—mail-ordering, in his little girls' name, a "dewat" Tommygun and a brand-new barrel.[23] When he had the combination working he called in the photographers. Fortunately for the editor, justice triumphed over the letter of the law, for he did not go to jail for the unlicensed manufacture of a fully automatic weapon.

On July 1, 1958, the Treasury Department discontinued its DEWAT program and resumed gun-to-owner registration. This curtailed the "rewatting" hobby, or at least restricted it to guns sold under the war trophy program.

Two years later the Army decided against disposing of large stocks of obsolete weapons on the commercial surplus market for fear of aggravating the country's perennial "gun problem," not to mention the Treasury Department. Accordingly, it sold them as metal scrap after each gun had been "demilitarized" by cutting torch or other brutal means. But some of the Army torchmen had no more heart for their work than had the gun-shop welders earlier. Of the thousands of Thompsons and M3s sold to several junkyards in the early 1960s, a significant number needed little more than patching, plus a new barrel and general overhaul.

Needless to say, news of this gold mine soon reached eager gunbuffs, who bought them by the pound or by the ton. Before the Treasury Department put a stop to this, a few thousand "demils," most of them Thompsons, had been snapped up by both collectors looking for bargains and by right-wing extremists looking for firepower.

In the spring of 1964 Richard Lauchli, leader of some Illinois patriots who viewed even the Minutemen as soft on Communism, attracted Treasury attention by running the following advertisement in *Shotgun News*, a paper published for gun fanciers:

> RIGHTIST & Guerilla Warfare Groups: If you are a recognized and responsible group (burden of proof is on you), then I have a special grade of Thompson 1928A1 "Demil" available to your group at regular price. Absolutely no sale of these to individuals. . . .[24]

In December 1964 Lauchli was sentenced to thirty months in prison for selling a truckload of reactivated Thompsons and other weapons to a group of Treasury agents posing as responsible revolutionaries. After losing several appeals and serving most of his sentence, Lauchli was paroled in February 1969. Less than two months later, on April 17, Treasury agents raided his Collinsville home and a nearby farm where they seized several hundred Thompsons, plus other ordnance, and Lauchli. The submachine guns were unrestored "demils" left over from the original batch, but they qualified as unregistered machine guns under new provisions of the Gun Control Act of 1968.

While Numrich wants no part of the deactivated gun market, he would not mind indulging the romantic whims of American gun buffs who hanker to own a live and legal Thompson. With Treasury Department approval, he has designed a semiautomatic Thompson which he hopes to one day market for about $180, "if we ever find the time." Designated the Model 1927A2, for the sake of tradition, the gun is essentially an M1A1 Thompson with a sixteen-inch barrel (the legal minimum), pistol foregrip, and provision for drum magazine. Unlike the original Model 1927 Thompson "carbine," the Numrich A2 model cannot be converted to fire full automatic by substituting a few parts. According to Numrich, it would be harder ("since nothing is impossible") to turn it into a submachine gun than it would be to so modify most semiautomatic rifles and pistols now on the market.

Thanks to movies and television, to increasing popular nostalgia for the 1920–40 period, and to its present scarcity, the Thompson has started taking on the quality of a genuine antique. Most privately owned Tommyguns, dead or live, are now in the hands of legitimate collectors, and there is an increasing interest in replicas. About 1960 Edmond H. de la Garrique, an expert machinist then living in Rochester, Michigan, built a number of scale-model Thompsons that have acquired a collectors' value of their own. The guns were about half the size of the real submachine gun, but perfect in every mechanical detail except actual firing features. The miniatures had no firing pin, extractor, or feed ramp; the barrels were bored and rifled for .22 and .25 caliber, but not chambered. Thirty-three were made in Models 1928 and 1928A1, and about fifty in Models M1 and M1A1. Each was carefully finished, properly marked, and either blued or Parkerized, as appropriate to the particular model. The walnut stocks also were perfect reproductions. De la Garrique's hopes of marketing the models for collectors ran into financial troubles, however, and he eventually abandoned the project. Most of the finished models did eventually go to collectors, for as much as $800, and at least one person yielded to the temptation to try to make them work. He succeeded to the extent of getting some eleven of the models confiscated and destroyed by the Treasury Department.[25]

About 1963 a Houston, Texas, firm built, and sold "dummy" M1928A1 Thompsons as wall decorations. The guns were cast in aluminum from a mold made from a real Thompson, and had no moving parts. But considerable realism was achieved by equipping the black-finished castings with surplus Thompson stocks, magazines, and Lyman rear sights.

In 1969 two firms were planning to market replica Thompsons made of cast zinc, built by ingenious Japanese toymakers. Accurate in almost every detail, the guns not only "fire" full-automatic, but eject a stream of smoking shells in the finest Tommygun tradition. The dummy .45 ACP cartridges have a bored-out nose in which paper caps are inserted. The bolt chambers a cartridge from the magazine in the usual manner, and an obstruction in the barrel fires the cap, which creates enough pressure to blow back the bolt. One Tokyo firm has built working "prototypes" of the guns in both the Model 1921 and M1A1 and is supposed to eventually manufacture them for William B. Edwards' Gold Rush Gun Shops in San Francisco and Afton, Virginia. Another Japanese firm is already planning production of a similar cap-firing replica Model 1921 Thompson to sell for around

$50. Tom Nelson of Hunters Lodge and Interarms Corporation describes this model as a "complete facsimile." It is to be marketed in late 1969 by Replica Models, Incorporated, of Alexandria, Virginia.

As it becomes more and more a valuable antique firearm, any modern bank robber would do well to sell his Thompson, pick up a shotgun and a hacksaw, and pocket a few hundred dollars' difference. Now and then the Tommygun still makes crime news, however. One of the more spectacular incidents occurred in 1961 when one Bobby Randall Wilcoxson robbed a Brooklyn bank with a "chopper" in the true John Dillinger style, killing a guard and earning himself top billing on the FBI's "Ten Most Wanted" list. Wilcoxson's partner, Albert Nussbaum, had bought the gun from a part-time gun dealer who had acquired it from a gunsmith. The gunsmith had gotten it from a friend who had found it among some rubbish along the Niagara River in Tonawanda, New York, some eight years before. The gun was in bad condition, and neither the dealer nor the gunsmith had attempted to restore it. Nussbaum, however, cleaned it up, replaced several parts, and managed to get it back in firing order.[26]

Outside of the United States, especially in the smaller Latin American countries, the Tommygun still performs an occasional assassination. In Vietnam, NLF guerrillas have fought with Thompsons acquired from the Chinese, who got them from Chiang Kai-shek's Nationalists during and after World War II. Despite its current status as a military antique, the Thompson still is in service in most parts of the world, officially or otherwise.

But apart from occasional crimes, confiscations, and limited use by police, militia, and guerrillas in foreign countries, the Tommygun no longer plays any significant role as a firearm. In the United States it remains a common weapon of municipal police, county sheriffs, and the FBI, but mostly as a relic of the 1920s and 1930s. For all practical purposes the company that brought out the Thompson is defunct, its inventors are forgotten, and the gun itself is obsolete. It survives mainly in the popular imagination as an American cultural artifact of the Roaring Twenties and the Dillinger era, as seen on the Late Late Show.

Notes

Book and article references to John Thompson, John Blish, and the Tommygun, as cited in the footnotes, must be used with caution. Most writers on the subject have relied heavily on Auto-Ordnance promotional literature and catalogs, which contain much inaccuracy or misleading information. Early Auto-Ordnance advertisements and news releases sometimes confused the features of the 1921 production model with those of the prototypes or exaggerated later modifications to make different guns. Most historical accounts of the submachine gun's development—and of the roles played by John Thompson and John Blish—are based on a few sketchy biographies whose errors have been perpetuated and sometimes expanded by subsequent writers. Therefore, the information in a particular source should not be regarded as necessarily accurate, except for the fact to which a footnote specifically refers. Where sources disagree on facts, the author has cited the one that could be verified independently or seemed otherwise to be the most reliable. Probably the best single source of technical information on Thompson guns is Thomas Nelson's *The World's Submachine Guns.*

CHAPTER ONE

1. Unless otherwise noted, all biographical information on members of the Thompson family is from General John T. Thompson's personal scrapbook, loaned to the author by Mrs. Evelyn Adams (formerly Mrs. Marcellus H. Thompson) of Morristown, New Jersey. Many of the scrapbook's newspaper and magazine clippings are not dated or identified, but most of the same information can be found in shorter form in the following sources: *National Cyclopaedia of American Biography*, XXIX, 297–9 (entries for James, John, and Marcellus Thompson); the obituaries of Marcellus and John Thompson contained respectively in the 1940 and 1941 editions of the *Annual Report*, published by the United States Military Academy Association of Graduates, West Point, New York; and John Thompson's obituary in the *New York Times*, June 22, 1940. These sources all contain errors and must be used with caution.

2. Thompson scrapbook (Newport, Kentucky, newspaper clipping, *c.* 1900, reporting Thompson's return from duty in Cuba).
3. Thompson scrapbook (clipping from a military journal, *c.* 1901, summarizing Lt. John H. Parker's newly published book, *History of the Gatling Gun Detachment*).
4. Julian S. Hatcher, *Textbook of Pistols and Revolvers; Their Ammunition, Ballistics and Use* (Plantersville, South Carolina, 1935), 415 ff; Theodore H. Eickhoff to the author, twelve letters, 1963–1967, and interview with Eickhoff, April 1967 (see Acknowledgments).
5. Thompson scrapbook ("An Able Officer Promoted," clipping, source given as "Arms & The Man, Oct. 30, 1913").
6. *Ibid.*
7. M[arcellus] H. Thompson, *Notes on Present Day Automatic Arms Design in Relation to Their Tactical Employment* (booklet, n.p., 2nd ed., 1924), 3–8; "Machine-Guns," *American Mercury*, April 1928, 453–6.
8. M. H. Thompson, 6 (citing Longstaff and Atteridge, *Book of the Machine Gun*).
9. Quoted in "Machine-Guns," *American Mercury*, April 1928, 454.
10. J[ohn] T. Thompson, *Modern Weapons of War* (Davenport, Iowa, 1905), 74, 94–7 (paper read before the Davenport Contemporary Club and printed as a pamphlet).
11. W. H. B. Smith and Joseph E. Smith, *Small Arms of the World* (Harrisburg, Pennsylvania, 7th ed., rev., 1962), 102, 155.
12. Eickhoff letters.
13. Harold L. Peterson, *The Remington Historical Treasury of American Guns* (paperback ed., New York, 1966), 137–42.
14. Eickhoff letters.
15. Incorrectly reported as John N. Blish in some sources. About 1907 Commander Blish legally dropped his middle name altogether.
16. James K. Blish, *Genealogy of the Blish Family in America* (Kewanee, Illinois, 1905), 191–2; U.S. Naval Academy Library to the author, April 1964.
17. Eickhoff letters. See also: Arthur F. Macconochie, "The Thompson Sub-machine Gun," *Steel*, November 10, 1941, 66–70.
18. Quoted in P. H. Torrey, "The Thompson Submachine Gun," *Marine Corps Gazette*, June 1921, 143–5.
19. *Ibid.*
20. Quoted in Elton Atwater, *American Regulation of Arms Exports, 1941* (Washington, D.C., 1941), 8.
21. William C. Whitney, quoted in *Dictionary of American Biography*, VIII, Pt. 2, 266 (Ryan bio.).

22. *New York Times*, November 24, 1928 (Ryan's obituary).
23. *Ibid.; D.A.B.*, VIII, Pt. 2, 265–8.
24. *D.A.B.*, IV, Pt. 2, 272–3; *New York Times*, August 21 and 26, 1928 (Harvey's obituary and follow-up); Thompson scrapbook.
25. "An Ambassador to the Court of St. James's," *New Statesman* (London), May 7, 1921, 123–4.
26. H. I. Phillips, in his newspaper column "The Once Over" (clipping in Thompson scrapbook; source not given).
27. Willis Fletcher Johnson, *George Harvey: A Passionate Patriot* (Boston and New York, 1929), 297.
28. "America and Britain: A Clean Slate," *Spectator* (London), May 20, 1922, 612.
29. Personal records of Thomas A. Kane, New York, onetime attorney for the Thompson family. (Papers include preliminary prospectus and stock registration statement for Thompson Automatic Arms Corp., 1939; Auto-Ordnance Corp. *Annual Report* for 1941; and an untitled deposition dated September 28, 1944, containing corporate history of Auto-Ordnance.)
30. Eickhoff letters.
31. Biographical information on Theodore H. Eickhoff is from his letters and interview.
32. George E. Goll to the author, eleven letters, 1963–1968 (see Acknowledgments). Biographical information on Goll is from his letters.
33. A. R. Stanley, "Uncle Sam's Premier Gunman—Col. John T. Thompson," Philadelphia *Public Ledger Magazine Section*, July 28, 1918, 1.
34. Eickhoff interview.
35. *Ibid.*
36. Information on Auto-Ordnance development work during 1916–1918 is from the letters of Eickhoff and Goll; "Report of Work on the Development of the Blish Patent, Cleveland, Ohio, November 12, 1917" (a 31-page progress report furnished by Mr. Eickhoff); and V. B. Gray, "Death Rattles in these Guns," Cleveland *Plain Dealer* Sunday magazine section, October 24, 1920, 5.
37. Eickhoff interview.
38. *Ibid.*

CHAPTER TWO

1. Eickhoff interview.
2. Oscar V. Payne to the author, eight letters, February to May 1967

(see Acknowledgments). Biographical information on Mr. Payne is from his letters.

3. Payne letters.
4. *Ibid.*
5. Eickhoff interview.
6. *Ibid.*
7. *Ibid.*
8. Payne letters.
9. *National Cyclopaedia of American Biography*, XXIX, 299.
10. William Crozier, *Ordnance and the World War* (New York, 1920), 59, 65; Sevellon Brown, *The Story of Ordnance in the World War* (Washington, D.C., 1920); *N.C.A.B.*, XXIX, 299.
11. Thompson scrapbook.
12. Eickhoff letters.
13. Payne letters.
14. Historical background of the submachine gun and related weapons is from Smith and Smith; George M. Chinn, ed., *The Machine Gun; History, Evolution, and Development of Manual, Automatic, and Airborne Repeating Weapons* (Washington, D.C., 1951), I; R. K. Wilson, *Textbook of Automatic Pistols* (Plantersville, South Carolina, 1943; originally published in England *c.* 1937); and Thomas B. Nelson with Hans B. Lockhoven, *The World's Submachine Guns (Machine Pistols)*, I (Cologne, Germany, 1963).
15. Information on the experimental Maxim gun was obtained in March 1969 from Val Forgett, Jr., who has donated the gun to the Marine Corps Museum, Quantico, Virginia. Museum custodian Garvice Costner has furnished the following approximate measurements: weight, 10 lb., 6 oz.; over-all length, 35 in.; barrel length, 18 in.; receiver length, 11 in.; receiver width, 2 in.; length of chamber, 1-11/16 in., with little or no taper. The gun appears to feed from right to left by means of reciprocating feed block channeled to grip a rimmed cartridge throughout the chambering-firing-extraction cycle. A small hole in the feed block comes into line with the firing pin hole in the bolt face at the moment of breach closure, permitting the pin to reach the cartridge primer. As of this writing the gun has not been displayed by the museum and few persons know of its existence.
16. Unless otherwise noted, information on the prototype submachine guns is from "Report of Development Work Conducted by Auto-Ordnance Corporation, Cleveland, Ohio, November 12, 1919" (a 29-page progress report furnished by Mr. Eickhoff and cited hereafter as Auto-Ordnance 1919 "Report").
17. Payne letters.

18. J. T. Thompson, 74; Torrey, 146; A. B. Richeson, "Submachine Guns," *Army Ordnance*, November–December 1920, 135.
19. "Thompson Sub-Machine Gun," *Army and Navy Journal,* October 2, 1920, 120.
20. Richeson, 136.
21. Information on the staff and activities of Auto-Ordnance in Cleveland is from the letters of Eickhoff and Payne.
22. Auto-Ordnance 1919 "Report"; Payne letters; Eickhoff interview.
23. Information on John Thompson's inventiveness is from Payne letters and Eickhoff interview.
24. U.S.M.A. *Annual Report,* 1940, 277. Description of Marcellus Thompson is based on this source and the following: Mrs. Evelyn Adams (formerly Mrs. Marcellus Thompson) to the author, two letters, November 1967 and February 1968; Thompson scrapbook; Eickhoff interview; and *N.C.A.B.*, XXIX, 297–8.
25. Eickhoff interview.
26. Adams letters.
27. Thompson scrapbook.
28. Quoted in Richeson, 135.
29. Quoted in Torrey, 148–9.
30. Philip B. Sharpe, "Sub-Machine Gun Performance," *American Rifleman*, December, 1934, 22; "History of Maguire Industries, Inc., Bridgeport, Conn.," *Historical Report, Springfield Ordnance District* (*c.* 1945), 2, contained in *Springfield Ordnance District Contractor Histories*, C, Pt. 3, WWII Records Division, National Archives (cited hereafter as "History of Maguire Industries"); "The New Submachine Gun," *Army and Navy Journal*, January 1, 1921, 519; Goll letters.
31. Eickhoff interview.
32. Auto-Ordnance office memorandum dated February 2, 1923, furnished by Mr. Eickhoff; Matthew J. Hall (stockbroker connected with Auto-Ordnance in 1939) to the author, two letters, March 1964.
33. Payne letters.
34. Auto-Ordnance promotional literature furnished by Goll, Payne, and Eickhoff.
35. Some early prototype Thompsons fired at a cyclic rate of about 1,500 rounds per minute, others at closer to 1,000, and the M1921 production model at 800 to 900 r.p.m. However, magazine articles and even Auto-Ordnance promotional literature of the early Twenties often attributed the higher firing rates as well as other prototype features to the Model 1921. According to George Goll, the higher firing rates were achieved by changing the angle of the Blish lock and by care-

fully polishing and fitting the internal parts. Initially, Auto-Ordnance considered a high rate of fire to be a selling point and continually stressed this feature. When it proved undesirable, the company reduced the rate as much as possible and sometimes sought to conceal it by listing only the gun's much lower "actual" rate of fire, which included the time required for changing magazines. In the M1921, as well as later models, the Thompson's rate of fire tended to vary from one gun to the next, and was often somewhat higher than advertised.

36. Auto-Ordnance shipping records for April, May, and June 1921, contained in records of the Department of Justice, Record Group 60, File 52–505, National Archives (records and documents pertaining to the smuggling case covered in Chapter Three, hereafter cited as Justice Department records).
37. "Test of Thompson Submachine Gun," *Army and Navy Journal*, April 16, 1921, 911.
38. Statement of Frank B. Ochsenreiter to J. Edgar Hoover, June 18, 1921, in Justice Department records.

CHAPTER THREE

1. *New York Times*, June 16, 1921.
2. "Sinn Fein and the Admiral," *Literary Digest*, June 25, 1921, 7–9.
3. *Chicago Tribune*, June 18, 1921; Justice Department records.
4. McHenry to Crim, memo, October 4, 1921, in Justice Department records.
5. Raymond Koontz (former Auto-Ordnance executive) to the author, six letters, 1963 to 1968.
6. J. Bowyer Bell to the author, five letters, July to December 1967 (see Acknowledgments).
7. McHenry to Crim, memo, October 4, 1921, in Justice Department records.
8. *Ibid.*
9. Bell letters.
10. *Ibid.*
11. Tom Barry, *Guerilla Days in Ireland* (2nd ed., Cork, 1955), 189–90.
12. *New York Times*, June 18, 1921.
13. Statement of Frank B. Ochsenreiter to J. Edgar Hoover, June 18, 1921, in Justice Department records.
14. Piaras Béaslaí, *Michael Collins and the Making of a New Ireland* (Dublin, 1926), I, 45, 62. Unless otherwise noted, other facts con-

cerning de Lacy are from Justice Department records. There may also have been a real Frank Williams (alias Frank Kerman, Lawrence Pierce, etc.) with whom de Lacy was associated. In any case, de Lacy succeeded in thoroughly and permanently confusing both newspaper reporters and the Department of Justice.

15. American "Sinn Feiners" spent more time and effort quarreling among themselves than working for the Cause. Far more useful to the I.R.A. were Irish exiles, like de Lacy, who were in the United States "on the run" and who worked closely with I.R.A. purchasing agents in securing and smuggling munitions. However, some of the Irish–American organizations did provide substantial sums of money.
16. The smuggling attempt has been reconstructed from news accounts published in the *New York Times* (June 16–22 and September 27–28, 1921; June 20, 1922) and the New York *World* (June 16–17, 1921; January 9 and June 20, 1922), and from Justice Department records, with reliance on Justice Department records where accounts differ in details.
17. *New York Times*, June 17, 1921.
18. Shortly before he left for Ireland, de Lacy was "politely kidnaped" by two Japanese. After treating him to a sumptuous meal they asked him to furnish them with Thompson submachine gun blueprints, to which he apparently had access through friends at Auto-Ordnance. De Lacy, somewhat mystified, obtained the blue prints and went his own way without asking any questions. (Bell letters.)
19. Telegram in Justice Department records.
20. *Ibid.*
21. Referred to in Crim to U.S. Attorney General, memo, October 28, 1921, in Justice Department records.
22. *Ibid.*
23. *New York Times*, June 20, 1922.
24. "Chopper," *Time*, June 26, 1939, 67.
25. Justice Department records.
26. Department of Justice to the author, April 1964.
27. *Ibid.*
28. Justice Department records; Bell letters.
29. *New York Times*, February 12 and June 20, 1922.
30. Florence O'Donoghue, *No Other Law* (Dublin, 1954), 250–52.
31. *New York Times*, August 2, 1922.
32. Bell letters. According to former I.R.A. officers Florence O'Donoghue and Joseph O'Connor, interviewed in Ireland in 1967 by Dr. Bell, some thirty Thompsons were smuggled into Cork in several overstuffed chairs and a sofa that were unloaded from a ship sometime before April 26, 1921. See also: Béaslaí, II, 184, 215.

33. Frank O'Connor, *Death in Dublin* (Garden City, New York, 1937), 238.
34. From "The Jolly Ploughboy," *Songs and Recitations of Ireland* (paperback, National Publications Committee, Cork, *c.* 1961), 19.

CHAPTER FOUR

1. Goll letters; H[ugh] B. C. Pollard, *A History of Firearms* (London, 1926, reprinted 1930), 247.
2. *Ibid.*
3. Marcellus Thompson to Morgan, May 3, 1921, in Justice Department records.
4. Marcellus Thompson to Morgan, April 26, 1921, in Justice Department records.
5. Merkling to Morgan, letter, June 9, 1921, in Justice Department records.
6. To the best of George Goll's recollections, only five Model 1923 Thompson guns were built, none were sold, and all were eventually dismantled. One country which showed some interest in the gun was the Soviet Union, but the "Bolshevist threat" of the 1920s ruled out the Russians as customers.

   The Model 1923 Thompson had an over-all length of thirty-six inches and weighed eleven pounds. Its optional accessories included a Maxim silencer, flash hider, bayonet, blank cartridge firing device, and sling with swivel mounts, and it was offered for sale, in minimum lots of ten, in five different cartridge sizes: .45 Remington–Thompson, .45 ACP, .351 Winchester, 9mm Mauser, and 9mm Luger.

   The .45 Remington–Thompson cartridge had a case length of 1.015 inches, a 250-grain jacketed bullet, a muzzle velocity of 1,450 feet per second, a muzzle energy of 1050 foot-pounds, and a claimed effective range of 650 yards. Despite its greater power, the Remington–Thompson reportedly was less accurate than the standard .45 ACP. Remington listed the cartridge from 1924 until about 1931, although it apparently was developed about 1922. Sources:

   Remington Arms Company to the author, April, 1968; Auto-Ordnance catalog for 1923; A. A. Blagonravov, *Small Arms Matériel* (Moscow, 1945); H. P. White, B. D. Munhall, and Ray Bearse, *Centerfire Pistol and Revolver Cartridges* (Rev. ed., New York and London, 1967), 96; Sharpe, 22; Goll letters.
7. Auto-Ordnance promotional literature. The Peters–Thompson shot cartridge was introduced about 1922 especially for the Tommygun. Its over-all length was slightly greater than the .45 ACP, and it

required a special eighteen-shot magazine which Auto-Ordnance designated the Type XVIII. According to Auto-Ordnance promotional literature, the shot cartridge was effective "at about the usual shotgun ranges," but in fact the barrel riflings dispersed the shot over a wide area at ranges beyond fifteen or twenty feet. About 1922 or 1923 Remington introduced another special cartridge for the riot market. Known variously as the .45 ACP Slug cartridge, Pistol Riot cartridge, and Thompson Riot Cartridge, the projectile consisted of a hollow copper alloy jacket enclosing a lead ball and three lead discs. Theoretically the components separated in flight and quelled as many as five rioters per shot, but the accuracy proved extremely poor and the components tended to cling together. Auto-Ordnance promoted the shot cartridge vigorously, but never recommended the other. (Remington letter; Sharpe, 22; White, *et al.*, *Centerfire Revolver Cartridges*, 96.)

8. E. C. Crossman, "A Pocket Machine Gun," *Scientific American*, October 16, 1920, 404, 413–14.
9. *New York Times*, May 11, 1922.
10. *New York Herald*, May 11, 1922.
11. *Ibid.*, May 1, 1922.
12. Auto-Ordnance catalog for 1923.
13. "Thompson Sub-Machine Gun," *Army and Navy Journal*, October 2, 1920, 120.
14. *Ibid.*
15. Richeson, 134.
16. Auto-Ordnance advertising leaflet, *c.* 1921.
17. H. G. Hartney, "The New Larsen All-Metal Attack Plane," *Army and Navy Journal*, November 5, 1921, 218, 221, 225; "Science and Invention" (clipping from unidentified military journal, furnished by Oscar Payne).
18. Hartney, 221.
19. Ray Wagner, *American Combat Planes* (Garden City, New York, 1960), 54; Eickhoff interview.
20. "The New Submachine Gun," *Army and Navy Journal*, January 1, 1921, 519.

CHAPTER FIVE

1. *New York Herald*, May 11, 1922.
2. Hugh B. C. Pollard, "Gun Running and the Traffic in Arms," *Saturday Evening Post*, November 24, 1923, 110.

3. Robert Carse, *Rum Row* (New York, 1959), 91–2; Herbert Asbury, *The Great Illusion: An Informal History of Prohibition* (Garden City, New York, 1950), 245, 250–51.
4. Information on Prohibition crime and gangland slayings is from the following sources: Asbury; Fred D. Pasley, *Al Capone: The Biography of a Self-Made Man* (New York, 1930); John H. Lyle, *The Dry and Lawless Years* (Englewood Cliffs, New Jersey, 1960); Kenneth Allsop, *The Bootleggers and Their Era* (Garden City, New York, 1961); and contemporary newspaper stories.
5. Illinois Association for Criminal Justice, *The Illinois Crime Survey* (Chicago, 1929), 923 (hereafter cited as *Illiinois Crime Survey*).
6. *Chicago Herald & Examiner*, September 26, 1925.
7. Chicago *Daily News*, October 5, 1925.
8. *Chicago Tribune*, February 10, 1926.
9. *Ibid.*, February 11, 1926.
10. *Illinois Crime Survey*, 831.
11. *Chicago Tribune*, May 5, 1926.
12. Payne and Goll letters.
13. William G. Shepherd, "Machine Gun Madness," *Collier's*, December 11, 1926, 8.
14. *Chicago Tribune*, May 2, 1926.
15. Francis Russell, "Sacco Guilty, Vanzetti Innocent?", *American Heritage*, June 1962, 2 ff.
16. Information on New York crime and criminals is from contemporary accounts in the *New York Times.*
17. *New York Times*, October 16 and 27, 1926.
18. *Ibid.*, March 8, 1930.
19. *Ibid.*, January 4, 1929.
20. *Ibid.*
21. *Ibid.*, January 6, 1929.
22. *Ibid.*, July 29, 1931.
23. *Ibid.*, February 8, 1932.
24. *Ibid.*, October 24, 1935.

CHAPTER SIX

1. Roger Touhy with Ray Brennan, *The Stolen Years* (Cleveland, 1959), 72.
2. William Piznak, "Present-Day Tommy Guns," *American Rifleman*, January 1951, 30. This article gives no date with the quote, but if

the British actually rejected the Thompson for this reason, it was probably in 1928, the year Britain conducted extensive small arms tests.

3. Quoted in Frederick L. Collins, *The FBI in Peace and War* (New York, 1943), 13. Information on the bank robbers and kidnapers of the 1930s is from the following sources: Collins; Don Whitehead, *The FBI Story* (New York, 1956); John Toland, *The Dillinger Days* (New York, 1963); Courtney Ryley Cooper, *Ten Thousand Public Enemies* (New York, 1935); Herbert Corey, *Farewell, Mr. Gangster!* (New York, 1936); J. Edgar Hoover, *Persons in Hiding* (Boston, 1938); Melvin Purvis, *American Agent* (Garden City, New York, 1936); and contemporary newspaper accounts.
4. Quoted in Andrew Sinclair, *Era of Excess: A Social History of the Prohibition Movement* (Harper Colophon paperback ed., New York, 1964), 220.
5. Auto-Ordnance catalog for 1936.
6. Max Lowenthal, *The Federal Bureau of Investigation* (New York, 1950), 416.
7. "Crime: Cummings on Warpath; J. E. Hoover Hits at Political Pull; Conference Acts Against 'Vermin,'" *Newsweek*, December 22, 1934, 5.
8. "Federal" was not officially added to the bureau's name until 1935. Prior to that time it was the Bureau (or Division) of Investigation of the Department of Justice. For the sake of convenience, this book will use the familiar term FBI regardless of the date of reference.
9. Collins, 15.
10. From "Pretty Boy Floyd" in Alan Lomax, *The Folk Songs of North America* (Garden City, New York, 1960), 427, 437.
11. "Oklahoma's 'Bandit King,'" *Literary Digest*, December 10, 1932, 26–7.
12. July 1, 1933, 34.
13. *New York Times*, July 14, 1933.
14. *Ibid.*, September 12, 1933.
15. *Ibid.*, July 24, 1933.
16. *Ibid.*, September 27, 1933.
17. Hoover, *Persons in Hiding*, 149.
18. Quoted in Emma Parker and Nell Barrow Cowan, *Fugitives: The Story of Clyde Barrow and Bonnie Parker*, compiled, arranged, and edited by Jan I. Fortune (Dallas, 1934), 242–5.
19. *Austin-American* (Austin, Tex.), August 21, 1959.
20. In his book *The Dillinger Days*, John Toland presents evidence that Dillinger also had a real pistol which had been smuggled to him in his cell by a bribed Indiana judge.

21. Corey, 69.
22. J. Edgar Hoover with Courtney Ryley Cooper, "The Real Public Enemy No. 1," *American Magazine*, April 1936, 16–17.
23. Address by J. Edgar Hoover in *Proceedings of the Attorney General's Conference on Crime, 1934* (Washington, D.C., 1936), 32.
24. *New York Times*, August 24, 1934.
25. *Ibid.*, May 11, 1922.
26. Owen P. White, "Machine Guns for Sale," *Collier's*, December 4, 1926, 11.
27. Edgar Sisson, "Making Massacre Easy: Super-Arms for Our Gangsters," *Today*, May 19, 1934, 3.
28. *New York Times*, September 6 and 7, 1928.
29. Owen P. White; see n. 26.
30. J. E. Thompson Sturm, "First Exposé of Our National Scandal: Illicit Machine-Gun Traffic That Revolutionized Gangsterism," New York *Daily Mirror* magazine section, August 14, 1932 (feature published in three parts, on August 7, 14, and 21, 1932, hereafter cited by author and date).
31. Shepherd, 36.
32. Address by J. Weston Allen in *Proceedings of the Attorney General's Conference on Crime*, 1934, 254.
33. Sisson, 4.
34. Shepherd, 8, 36.
35. "History of Maguire Industries," 4.
36. Corey, 32.
37. *Ibid.*, 44.
38. *San Antonio Express*, May 1 and 16, 1934.
39. Sisson, 3.
40. Sturm, August 7, 1932.
41. J. Weston Allen, *loc. cit.*, 255, 257.
42. *Ibid.*, 256–7.
43. *New York Times*, April 24 and 25, 1934; see also Toland, 286–7.
44. Sturm, August 21, 1932.
45. U.S. Senate, Subcommittee of the Committee on Commerce, *Crime and Crime Control, Investigation of So-Called Rackets, Hearings*, Pursuant to S.R. 74 (73rd Cong., 2nd Sess., 1933–34), I, Pt. 3, 287 (hereafter cited as Senate Rackets Committee hearings).
46. *Ibid.*, 293.
47. *Ibid.*, 318.
48. *New York Times*, December 28, 1926; Sturm, August 21, 1932.
49. Senate Rackets Committee hearings, I, Pt. 3, 294–5.
50. New York Police Department to the author, April 1964.

51. "G-men: New York Battle Revives Talk of a Hoover Resignation," *Newsweek*, December 26, 1936, 16.

CHAPTER SEVEN

1. Auto-Ordnance memo, February 2, 1923, in Eickhoff records.
2. U.S. Coast Guard Academy Library to the author, May 1964; U.S. Coast Guard Public Information Division to the author, May 1964.
3. "Weapons Used by Marines to Be Shown in New Poster," *Recruiters' Bulletin* [USMC], September 1921, 5; "Professional Notes," *Marine Corps Gazette*, December 1925, 203.
4. "Professional Notes," *Marine Corps Gazette*, December 1927, 249.
5. Sandino made peace with the government in 1933 but retained a small armed following. A short time later he was kidnaped and murdered on the orders of Anastacio Somoza, a high-ranking Guardia Nacional officer who emerged as dictator of Nicaragua in 1936 and remained in power until his assassination in 1956.
6. *Thompson Submachine Gun, Model of 1928*, training booklet, supplement to *Weapons and Musketry*, USMC Correspondence School Quantico, Virginia, *c.* 1935, 25 (cited hereafter as USMC Correspondence School training booklet).
7. *Ibid.*
8. "Professional Notes," *Marine Corps Gazette*, December 1927, 249; June 1928, 148; USMC to the author, January, 1968.
9. Roger W. Peard, "Practical Employment of the Thompson Submachine Gun," *Marine Corps Gazette*, June 1930, 128.
10. USMC Correspondence School training booklet, 27.
11. U.S. Navy, Bureau of Naval Weapons, to the author, two letters, April and August 1964.
12. *Ibid.*
13. Payne letters.
14. Julian S. Hatcher, *Hatcher's Notebook* (Harrisburg, Pennsylvania, 2nd. ed., 1957), 266; Gen. Richard M. Cutts, Jr., to the author, September 1964.
15. "Marine Corps Mail Guards Carry Improved Machine Gun," *Leatherneck*, December 1926, 44.
16. *Handbook of the Thompson Submachine Gun* (4th ed., January 1929).
17. Goll letters.
18. *Ibid.*

19. *Ibid.*; Morgan to Merkling, May 17, 1921, in Justice Department records.
20. Nelson and Lockhoven, 55.
21. *Ibid.*, 54; Goll letters.
22. Unless otherwise noted, information on the Thompson Autorifle is from the following sources: Auto-Ordnance 1919 "Report"; "Test of Thompson Autorifle," *Army and Navy Journal*, October 27, 1921, 176; Herbert O'Leary, "Experimental Work at Springfield Armory," *Army Ordnance*, July–August 1928, 17–27; "Semi-Automatic Rifles, 1925–40," in *Reports of Tests of Rifles and Rifle Parts, 1925–43*, Office of the Chief of Ordnance, Industrial Service, Small Arms Division, Engineering Branch, Folder, Box 6, Entry 846, Record Group 156, National Archives; Hatcher, *Hatcher's Notebook*, 46, 153; letters of Goll and Payne.
23. *Springfield Union*, April 25, 1926 (clipping in Thompson scrapbook).
24. *New York Times*, May 2, 1928. The Thompson Autorifle was built in several designs, in at least three calibers (.30, .276, and .303), and in several models, some of which differed only slightly. At least one may have been built to fire full-automatic. No published sources examine the Thompson Autorifles in any detail or offer any explanation of why they seem to have performed well in some tests and poorly in others. In 1949 two of the Model 1923 rifles were subjected to extensive study by a private ordnance testing laboratory. They were found heavy by World War II standards (10.1 and 12.1 pounds respectively, due to different stocks and magazines), extremely temperamental, and having many poor design features, including the need for lubricating pads in the magazines. (H. P. White Company, "Report of the Examination & Testing of the Thompson Automatic Rifles, Cal. .30, Model of 1923, On Behalf of the Kilgore Manufacturing Co., June 18, 1949," copy furnished by the H. P. White Laboratory of Bel Air, Maryland.)
25. Deposition in Kane records; Hall letters.
26. Registration statement in Kane records.
27. Deposition in Kane records; Cutts letter.
28. Preliminary prospectus in Kane records.
29. Quoted in Sturm, August 7, 1932.
30. Copy furnished by Thomas B. Nelson.
31. U.S. House, Committee on Ways and Means, *National Firearms Act, Hearings*, Pursuant to H.R. 9066 (73rd Cong., 2nd Sess., 1934), 66–7; Cutts letter.
32. Shepherd, 36.
33. *New York Times*, January 26, 1928. Auto-Ordnance was not com-

peting with shotgun manufacturers for the gangster market as the newspaper article seemed to suggest, but for sales to banks, industrial firms, and private protective agencies.

34. U.S. Senate, Subcommittee of the Committee on Education and Labor, *Violations of Free Speech and Rights of Labor, Hearings*, Pursuant to S.R. 226 (74–76th Cong., 1936–44), Pt. 7, 2609 (hereafter cited as La Follette Committee hearings); U.S. Senate, Special Committee to Investigate the Munitions Industry, *Hearings*, Pursuant to S.R. 206 (73–74th Cong., 1934–43), Pt. 7, 1781 (hereafter cited as Nye Committee hearings); deposition in Kane records.
35. *New York Times*, March 5, 1937.
36. La Follette Committee hearings, Pt. 1, 265–8; Pt. 7, 2486.
37. *Ibid.*, Pt. 7, 2491–5.
38. *Ibid.*, Pt. 15-D, 7005–7006.
39. *New York Times*, September 19, 1934.
40. Nye Committee hearings, Pt. 7, 1869–71.
41. *Ibid.*
42. *New York Times*, September 20, 1934.
43. National Council for the Prevention of War, *Now It Can Be Proved: Munitions Makers Indicted by Their Own Words* (booklet, Washington, D.C., *c.* 1935), 3; Nye Committee hearings, Pt. 38, 13, 221.
44. *New York Times*, August 9, 1933.
45. Philip Mohun, "The Revolution Racket," *Liberty*, July 21, 1934, 36; *New York Times*, November 6, 1935.
46. Sean O'Callaghan, *The Easter Lilly* (London, 1956), 195, 198; Ireland, Department of Defense, to the author, August 1964; Bell letters.
47. Nye Committee hearings, Pt. 5, 1197–8.
48. *Ibid.*
49. *Ibid.*, Pt. 7, 1867.
50. *Ibid.*, Pt. 7, 1867–8.
51. *Ibid.*
52. *Look* Magazine Editors, *The Story of the FBI* (New York, 1947), 184.
53. *New York Times*, December 25, 1934, and November 6, 1936.
54. Information on foreign submachine guns is from Nelson and Lockhoven and from Smith and Smith.
55. La Follette Committee hearings, Pt. 2, 638–40.
56. Nye Committee hearings, Pt. 7, 1893.
57. R. F. Sedgley, a Philadelphia arms dealer whose shady operations in the 1930s earned him the nickname "Sidewalk," where he reputedly did most of his business.

58. Nye Committee hearings, Pt. 7, 1903–1904.
59. About 1932 China sought to purchase a large number of Thompsons from Auto-Ordnance, but the sale apparently ran into financial obstacles and did not go through. The Chinese probably acquired a few Thompsons during the Twenties and early Thirties, but did not acquire large numbers until they were made available by the United States Government through Lend-Lease. Prior to World War II, most of the Tommyguns used by Chinese troops were copies manufactured by at least two Chinese arsenals from the late Twenties through the early Forties. These differed from the real Thompson only in workmanship and such external features as stock shape, sling attachment, number of cooling fins, and markings. A specimen at the USMC Museum at Quantico, Virginia, bears the serial number 17987. See Nelson and Lockhoven, 645, and *Submachine Guns*, III (U.S. Army Ordnance School, Aberdeen Proving Ground, Maryland, 1959), 38.

CHAPTER EIGHT

1. Unless otherwise noted, information on this topic is from the following sources: Thompson scrapbook; interview with Thomas A. Kane, April 1967; Adams letters; *New York Times*, April 4, 1937; and letters of John Thompson, furnished by Eickhoff and Payne.
2. Juliet Thompson's obituary, clipping furnished by Mr. Payne.
3. Thompson to Payne, September 7, 1930.
4. Thompson to Payne, August 16, 1931.
5. Eickhoff interview.
6. Thompson to Eickhoff, September 11, 1928.
7. *Worcester* (Massachusetts) *Evening Gazette*, January 16, 1931, commenting on an editorial in the *Chicago Tribune*.
8. See "Report of Ordnance Tests of Thompson Submachine Gun, by the Department of Experiment, The Infantry School, Fort Benning, Georgia," contained in records of the Adjutant General's Office, Record Group 94, Central Decimal Correspondence File 472.5 (Bulky Package No. 1574), National Archives.
9. *Ibid.*, 40–41.
10. From an untitled Army document quoting "OCM Item 4007, 1924, Pages 3874–3879," copy furnished by William B. Edwards. These same findings are reported in greater detail in *Ibid.*
11. *Ibid.* (Letter from the Office of the Chief of Infantry to Adjutant General of the Army, January 29, 1923.)

12. Ordnance Committee Minutes Nos. 7055 and 9627, contained in Records of the Ordnance Department, Record Group 156, National Archives (cited hereafter as OCM). Basis of issue was one submachine gun per armored vehicle and motorcycle of a mechanized cavalry regiment or armored car troop, "subject to the availability of funds."
13. Quoted in OCM No. 8338.
14. OCM Nos. 12655 and 14701. See also OCM Nos. 6821, 8025, 8338, 8495, 8667, and 9413.
15. Wilson, 303–304.
16. This account of the Thompsons' efforts to regain control of Auto-Ordnance is from the following sources: Kane interview and records, particularly the deposition; Koontz letters; Adams letters; and "History of Maguire Industries."
17. Statistics and figures from registration statement in Kane records.
18. Biographical information on Maguire is from *Bridgeport* (Connecticut) *Sunday Post*, September 2, 1945 (article based on an Associated Press feature by Richard Tompkins); and Irving Lieberman and Malcolm Logan, feature titled "Winchell's Pet Gazette: The Declining American Mercury," in *New York Post*, June 4, 1953.
19. *Bridgeport Sunday Post*, September 2, 1945.
20. Koontz letters.
21. Information concerning Maguire and Auto-Ordnance during 1939 and 1940 is from the following sources: "History of Submachine Guns, 1921 Through 1945," compiled by W. H. Davis and Capt. Andrew J. Gleason, Small Arms Division, Industrial Service, Ordnance Dept., Reel 64.342, in WWII Records Division, National Archives (hereafter cited by title); "History of Maguire Industries"; and two letters from the Savage Arms Corporation to the author, September and December 1963.
22. Herbert A. Stewart (former Savage official) to the author, January 1964; Goll letters.
23. "History of Submachine Guns, 1921 Through 1945," 4–6.
24. "History of Maguire Industries," 8, 11.
25. Kane interview.
26. Files of the *New York Times*.
27. Thompson to Eickhoff, undated, *c.* 1939.
28. Thompson to Eickhoff, July 27, 1939.
29. Thompson to Eickhoff, undated, *c.* 1939.

CHAPTER NINE

1. Unless otherwise noted, information on the establishment and operation of the Auto-Ordnance Bridgeport plant is from "Famous 'Tommy' Gun Made by Self-Contained Plant," *Iron Age*, July 2, 1942, 54–60.
2. *Bridgeport Sunday Post*, September 2, 1945.
3. Statistics from "History of Submachine Guns, 1921 Through 1945," 6, 11, Appendix; and "History of Maguire Industries," 13, 15.
4. "History of Maguire Industries," 13–14; *Bridgeport Sunday Post*, November 1, 1942.
5. Eickhoff interview.
6. Stewart letter.
7. "History of Submachine Guns, 1921 Through 1945," 10, Appendix; Savage letters. Because it was introduced in 1942, the M1 Thompson is sometimes referred to as the "42M1." Savage built only about eighty M1 Thompsons before switching over to production of the M1A1, but the Auto-Ordnance plant at Bridgeport produced some 285,000 of the M1 model.
8. *Ibid.*
9. Ray Bearse, "The Thompson Submachine Gun, Weapon of War and Peace," *Gun Digest* (21st ed., 1967), 52.
10. "History of Submachine Guns, 1921 Through 1945," 30. The thirty-round magazine is commonly called the Type XXX.
11. *Ibid.*, 51–2.
12. Nelson and Lockhoven, 489.
13. Savage letters.
14. Unless otherwise noted, all information on the Army's submachine gun development and testing program is from "History of Submachine Guns, 1921 Through 1945."
15. Nye Committee hearings, Pt. 8, 1931–5; Hyde obituary notice in *American Rifleman*, February 1964, 81.
16. René R. Studler, "The New Submachine Gun, M3," *Army Ordnance*, September–October 1943, 349–50. Includes a discussion of Army small-arms research and development policies.
17. *Ibid.*
18. "History of Submachine Guns, 1921 Through 1945," 13.
19. *Ibid.*, 12–13.
20. Quoted in Nelson and Lockhoven, 490.
21. *Small Arms and Small Arms Ammunition*, II, Book I of *Record of Army Ordnance Research and Development*, Office of the Chief of

Ordnance, Research and Development Service (Washington, D.C., January 1946), Ch. 3 ("Submachine Guns"); "History of Submachine Guns, 1921 Through 1945," 16–18. In its early efforts to develop a light, compact weapon for paratroops, the Army conducted experiments with the M1911 Colt automatic pistol modified to accept a shoulder stock, a 9.5-inch barrel with cooling jacket, and twenty-round magazine. This was merely a refinement on the Colt pistol with large-capacity magazine used by some Allied fliers to shoot at enemy aircraft at the start of World War I, and the idea was soon abandoned. In 1944 the Army issued a "paratroop" version of the .30-caliber M1 carbine modified to fire full-automatic by means of a selector switch. Designated the U.S. Carbine, Caliber .30, M2, it was commonly equipped with a folding wire stock and differed from a "submachine" gun mainly in that it fired standard carbine ammunition. In some references this weapon is confused with the .45-caliber M2 submachine gun.

22. U.S. Army Matériel Command to the author, December 1968, and "History of Submachine Guns, 1921 Through 1945," 59, Appendix. From July 1, 1940, through August 31, 1945, the Ordnance Department procured a total of 2,033,267 Thompson and M3 submachine guns (including 25,000 M3s built in 9mm for the Office of Strategic Services, to be dropped to the "underground" in Europe), plus several thousand Reising submachine guns which were never distributed. The following chart shows procurement of .45-caliber Thompsons and M3s by year:

	M1928A1	M1	M1A1	M3	M3A1
1940	3,630	0	0	0	0
1941	213,790	0	0	0	0
1942	344,521	249,420	8,552	0	0
1943	570	36,060	526,500	85,100	0
1944	0	0	4,091	343,372	0
1945	0	0	0	177,192	15,469
	562,511	285,480	539,143	605,664	15,469

These figures include submachine guns procured through the Army for the Navy, Marine Corps, and Lend-Lease, which are estimated to represent about a quarter of the total.

23. Frank O. Hough, *The Island War* (Philadelphia and New York, 1947), 102.

24. Most of these guns were supplied before the United States entered the war, while Russia and the British Commonwealth countries were

in the process of developing new submachine guns or putting them into production. Probably most of the nearly 300,000 Thompsons built during 1940–1941 went to Allied armies. According to legend, 200,000 or more were lost to German U-boats in the sinking of cargo ships.

25. Constance McLaughlin Green, Harry C. Thomson, and Peter C. Roots, *The Ordnance Department: Planning Munitions for War* (Washington, D.C., 1955), Ser. 5 in III, pt. 2 of the series, *United States Army in World War II*, 43.
26. H. Duncan Hall, *North American Supply* (London, 1955), 132; "History of Maguire Industries," 8, 11.
27. Navy letters.
28. Quoted in "History of Maguire Industries," 9–10.
29. Quoted in *Ibid.*, 12.

CHAPTER TEN

1. "History of Maguire Industries," 15; Auto-Ordnance Corp., *Annual Report to Stockholders, 1944*; Lieberman and Logan.
2. Deposition in Kane records; *New York Times*, September 18 and December 5, 1940, and March 5, 1941.
3. Dallas S. Townsend (attorney who represented Cutts and Hoover) to the author, January 1964.
4. Cutts letter; Lieberman and Logan.
5. Lieberman and Logan; Koontz letters.
6. William B. Ruger to the author, January 1968.
7. *Ibid.*
8. Roger Marsh, "The Tommygun," *American Rifleman*, July 1957, 45; Nelson and Lockhoven, 53–4.
9. Nelson and Lockhoven, 57; "History of Submachine guns, 1921 Through 1945," Appendix.
10. Ruger letter.
11. Calvin Goddard, "New Developments: Caliber .22 Conversion Unit for the Thompson Submachine Gun," *Army Ordnance*, January–February 1943, 120–23 (see also U.S. Patent 2,427,304).
12. Eugene Daniel Powers (former Auto-Ordnance official) to the author, May 1964.
13. Lieberman and Logan.
14. Quoted in *Ibid.*
15. "Maguire of the Mercury," *Newsweek*, August 25, 1952, 49.

16. *Bridgeport* (Connecticut) *Sunday Post*, September 2, 1945.
17. *Newsweek*, August 25, 1952, 49.
18. "Mencken's *Mercury* Now," *Newsweek*, March 14, 1960, 92; "Blowup at the Mercury," *Time*, October 3, 1955, 72; Lieberman and Logan.
19. In *The World's Submachine Guns*, Thomas Nelson reports the existence of a .45-caliber "Egyptian Thompson" manufactured experimentally in Egypt in the early nineteen-fifties. The gun somewhat resembles the Thompson in profile, uses Thompson magazines, and has a cylindrical receiver marked "Auto-Ordnance Corp., Bridgeport, Connecticut."

    Unless otherwise noted, information on Auto-Ordnance after 1945 is from the following sources: George R. Numrich, Jr., to the author, seven letters, 1963–1968; interview with Numrich, April 1967; Treasury Department to the author, January 1967; and Bob Zwirz, "Don't Junk It; Shoot It!", *Gun World*, June 1965, 41–5.
20. Numrich interview.
21. Jack Olsen, "The Big Itch They Call Little John," *Sports Illustrated*, January 11, 1965, 56.
22. John S. Tompkins, "The Honest Gunmakers and Their Counterfeit Carbines," *True*, August 1963, 86 ff.
23. Ashley Halsey, Jr., "Murder Weapons for Sale," *Saturday Evening Post*, February 8, 1958, 27 ff.
24. *Shotgun News*, April 15, 1964, 24.
25. Thomas S. Pendergast, Jr., letter and interview, April 1969.
26. FBI to the author, September 1963.

APPENDIX A

Handbook of the Thompson Submachine Gun

MODEL OF 1921

PREFACE

This handbook is issued to furnish such information as is necessary for the operation and care of the Thompson Submachine Gun.

It is urged that the directions and *precautions* contained herein be carefully noted before operating the gun.

TABLE OF CONTENTS

LIST OF COMPONENTS OF THOMPSON SUBMACHINE GUN

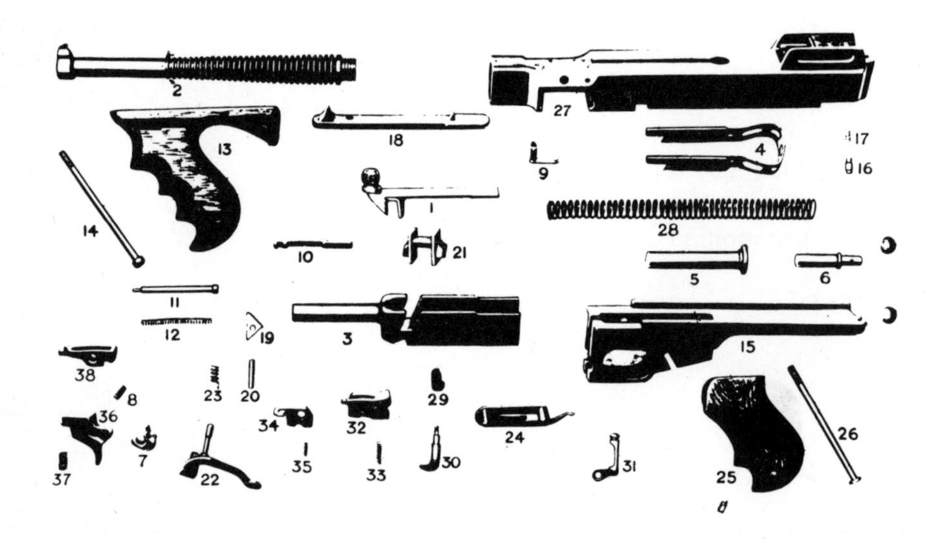

1. Actuator.
2. Barrel.
3. Bolt.
4. Breech Oiler. (including Felt Pads).
5. Buffer. (including Fiber Discs).
6. Buffer Pilot.
7. Disconnector.
8. Disconnector Spring.
9. Ejector.
10. Extractor.
11. Firing Pin.
12. Firing Pin Spring.
13. Fore Grip.
14. Fore Grip Screw.
15. Frame.
16. Frame Latch.
17. Frame Latch Spring.
18. Grip Mount.

19. Hammer.
20. Hammer Pin.
21. Lock.
22. Magazine Catch.
23. Magazine Catch Spring.
24. Pivot Plate.
25. Rear Grip.
26. Rear Grip Screw.
27. Receiver.
28. Recoil Spring.
29. Rocker.
30. Rocker Pivot.
31. Safety.
32. Sear.
33. Sear Spring.
34. Sear Lever.
35. Sear Lever Spring.
36. Trigger.
37. Trigger Spring.
38. Trip.

MAGAZINES

Box Magazine (20-cartridge).
Body.
Follower.
Spring.
Floor Plate.

Type L Magazine (50-cartridge drum).
Body, assembled.
Cover, assembled.
Rotor, assembled.
Winding Key, assembled.*
Body Clip.*

Type C Magazine (100-cartridge drum).
Body, assembled.
Cover, assembled.
Rotor, assembled.
Winding Key, assembled.*
Body Clip.*

SIGHTS.

Front Sight (assembled to barrel).
Rear Sight.
Eye piece.
Rear Sight Base (assembled to receiver).

* These parts are identical for both Type L and Type C Magazines.

Sight Base Pin.
Sight Plunger.
Sight Plunger Spring.
Sight Leaf (with slide retaining pin).
Sight Slide.
Sight Slide Catch.
Sight Slide Catch Screw.
Windage Screw, assembled.
  Consisting of: Windage Screw.
    Windage Screw Collar.
    Windage Screw Collar Pin.

APPENDAGES AND ACCESSORIES.

Butt Stock complete, consisting of:
  Butt Stock (stripped).
  Butt Stock Slide Group.
    Butt Stock Slide.
    Butt Stock Catch.
    Butt Stock Catch Pin.
    Butt Stock Catch Spring.
    Butt Stock Screw, large.
    Butt Stock Screw, small.
  Butt Plate Group.
    Butt Plate (stripped).
    Butt Plate Cap.
    Butt Plate Pin.
    Butt Plate Spring.
    Butt Plate Spring Screw.
    Butt Plate Screw, large.
    Butt Plate Screw, small.

Cleaning Rod.
Brass Wire Brush.
Gun Case.
Oil Can.

I. Description of Parts

The Thompson Submachine Gun, without sights and butt stock, consists of thirty-eight component parts, as follows:

The *actuator* (1) as the name implies, performs the function of actuating the lock under impulse of the recoil spring and also furnishes a projection for manual operation of the bolt mechanism. It has two fingers extending downwardly engaging the cross-bar of the lock; the rear finger also furnishing a seat for the recoil spring. The knob extending upward through the receiver furnishes a purchase for the fingers of the operator, and the rearwardly extending portion of the actuator closes the clearance slot in the receiver against dust and rain when the bolt is in closed position.

The *barrel* (2) is chambered and rifled for the standard caliber .45 automatic pistol ammunition. It has an overall length of 10½ inches and a bore length of approximately 9¾ inches. The rear portion of the barrel is provided with annular fins for cooling purposes.

The *bolt* (3) consists of a rectangular body portion and a forwardly projecting cylindrical portion of a diameter sufficient to support the face of the cartridge and furnish a supporting flange therefor. The rear rectangular portion is provided with a cavity for the recoil spring, with obliquely inclined slots for the lock, with notches for the sear, and with a seat for the hammer. The lower forward end of this portion of the bolt furnishes an abutment to seat the bolt in its forwardmost position against the receiver. The round portion of the bolt has a counterbored cartridge seat, a seat for the extractor, a clearance cut for the ejector, and carries centrally located a firing pin and firing pin spring.

The *breech oiler* (4) is formed of sheet metal to hold a felt pad on each side of the undercut portion of the receiver. It is held in place against movement by the buffer pilot.

The *buffer* (5) consists of a cylindrical tube closed at its forward end and provided with a flange at its rear end; the flange serving as a seat for the recoil spring, and as an abutment for the bolt. It contains a column of fiber discs which absorb the shock of recoil.

The *buffer pilot* (6) serves as a plunger in the buffer against the buffer discs and supports the buffer. A rearwardly extending projection fits loosely in a seat in the receiver and extends sufficiently beyond to aid in dismounting the gun.

The *disconnector* (7) consists of a body with a pivot extending therefrom and two lever arms projecting in a plane normal to the pivot; one arm engages with the sear lever and provides an abutment for the rocker; the other arm furnishes an abutment for the trip and limits the movement of the disconnector.

The *disconnector spring* (8) is seated in the trigger in a position to urge the disconnector to engage with the sear lever.

The *ejector* (9) consists of a threaded body portion which screws into the receiver, and a spring leaf by means of which the ejector can be screwed into place. On the under side the leaf is provided with a small stud, which when the ejector is assembled to the receiver engages with a depression in the receiver and locks the ejector in place. The lower end of the ejector body is shaped into a suitable form of ejector head.

The *extractor* (10) consists of a leaf spring body with a hooked head to engage the cartridge and a projecting stud to anchor in the bolt against axial movement. The rear portion of the body is provided with ribs which engage with an undercut groove in the bolt, to hold the extractor in position. The body of the extractor is given a slight curvature to keep the head under proper spring tension for engagement with the cartridge case, and for holding the extractor assembled to the bolt.

The *firing pin* (11) consists of a cylindrical body with its forward end reduced in diameter for striking the primers, and its rear end enlarged to furnish a seat for the firing pin spring, and to receive the blows from the hammer.

The *firing pin spring* (12) surrounds the firing pin and constantly urges the same rearwardly to prevent protrusion of the firing pin except at the moment of firing.

The *fore grip* (13) is a black walnut member which serves as a purchase on the gun for the left hand of the operator.

The *fore grip screw* (14) is of sufficient length to pass through the body of the fore grip and secure same to the fore grip mount.

The *frame* (15) serves as a mount for the receiver, contains all the firing mechanism, furnishes a seat for the rear grip and a support for the box magazine. The rearwardly extending projection is also utilized for attaching the butt stock.

The *frame latch* (16) is a small pin member pocketed in the rear end of the receiver, with its upper end reduced in diameter to serve as an arbor for the frame latch spring, and the lower end reduced to fit the frame. The latch locks the frame in assembled position to the receiver.

The *frame latch spring* (17) urges the frame latch in constant engagement with the frame.

The *grip mount* (18) furnishes a support to the fore grip on the receiver. It is secured to the under side of the front end of the receiver by means of an undercut, and held in place by a stud projecting upward behind the flange of the barrel. Its forward end is provided with a lug to contact with the barrel. The body is given a slight curvature upward before assembling, so that in assembled position the lug rests against the barrel. The grip mount is assembled to the receiver with the barrel.

The *hammer* (19) is a lever of triangular shape pivoted slightly below its center to obtain an increased length of upper arm. The lower arm protrudes sufficiently beyond the forward abutment of the bolt so that when the bolt in its forward movement approaches the receiver abutment, the hammer receives a blow against the receiver abutment and transmits this blow with increased velocity to the firing pin.

The *hammer pin* (20) serves as a pivot for the hammer.

The *lock* (21) is a member of titanium aluminum bronze with three groups of surfaces in three different planes. It resembles somewhat the letter "H" with lugs projecting from its sides. The center bar forms the body of the lock holding the two sides intact, and also serves as an engagement for the actuator. The surfaces thereof are inclined at an angle to create a downward pressure component on the lock resulting from the recoil spring pressure. The two side members engage the locking slots in the bolt, and the projecting lugs engage the locking grooves in the receiver. Relative to the bolt the movement of the lock is confined to an obliquely up and down movement in the bolt slots. Relative to the actuator the center bar moves obliquely up and down between the fingers of the actuator, but the angularity of the bolt-lock surfaces and the lock-actuator surfaces is such that a movement of the lock upward in the bolt causes the actuator to move slightly rearward, compressing the recoil spring and vice versa. Relative to the receiver, the lock at the stage of locking and unlocking moves obliquely up and down and after disengagement with the locking grooves the lock merely reciprocates in the receiver with the bolt. In the stage of locking, the lock follows the locking grooves in the receiver, but in so doing, due to the difference in angularity of the bolt-lock and the lock-receiver surfaces, it causes the bolt to travel forwardly and vice versa.

The *magazine catch* (22) consists of a pivot, a rearwardly extending lever for thumb control, and an arm extending forwardly and downwardly bifurcated and embracing a portion of the forward end of the trigger guard. The right member of this bifurcation is round and engages with the box magazine holding same in position on the gun. The left member of this bifurcation is rectangular and engages with the drum magazines, holding these in position on the gun.

The *magazine catch spring* (23) is a torsion spring surrounding the pivot of the magazine catch. It urges the magazine catch into engagement with the magazine.

The *pivot plate* (24) consists of a spring plate body on which are mounted two pins—one pin serving as a pivot for the trigger and trip—the other serving as a pivot for the sear and sear lever. Two fingers extend from the plate—the longer finger engages the safety, and the other engages the rocker pivot. These fingers serve to hold the rocker pivot and safety assembled to the frame; they also serve to hold these parts in their two designated positions—the safety in the "fire" and "safe" position, and the rocker in the "automatic" and "semi-automatic" position.

The *rear grip* (25) furnishes a purchase on the gun for the right hand of the operator. Its proximity to the trigger guard is such, that when gripping the rear grip the index finger is used conveniently for pulling the trigger, and the thumb is in position to operate the magazine catch, the rocker pivot and the safety.

The *rear grip screw* (26) passes through the length of the rear grip and secures the same to the frame.

The *receiver* (27) furnishes a skeleton for the gun and houses the entire bolt action. The front end is threaded to receive the barrel and its lower surface is undercut to furnish a seat for the grip mount. It is provided with a magazine clearance cut, in which are guideways for attaching the drum magazines. The funnel-shaped throat guides the cartridges from the magazine into the chamber of the barrel. An ejector opening is provided so that ejection will be upward to the right. A threaded hole furnishes a seat for the ejector. The cavity in the receiver in rear of the magazine clearance cut provides a housing and reciprocation space for the bolt mechanism. This cavity is provided at its forward end on each side with locking grooves which terminate in an undercut for reciprocal movement of the lock lugs. The rearmost position of this undercut is utilized to house the breech oiler. The upper surface of the receiver is slotted to provide a path for reciprocation of the actuator knob, and the rear end is provided with an axial hole to seat the buffer pilot and with a vertical hole to house the frame latch. The lower sides of the receiver are provided with undercut ways for securing the frame to the receiver.

The *recoil spring* (28) is supported at its rear end on the buffer. Its forward end rests in the cavity of the bolt and seats against the rear finger of the actuator. It does not only furnish force to drive the bolt forward and load a cartridge into the chamber, but by imparting inertia to the bolt in its forward movement it causes the firing of the cartridge by the hammer striking the receiver abutment. Since its pressure is applied to the bolt through the actuator and lock, it also urges the lock into locked position.

The *rocker* (29) is a vertical lever pivoted at its lower end on an eccentric portion of the rocker pivot; its upper end engages with the bolt and its forward edge contacts with the disconnector.

The *rocker pivot* (30) consists of a body of two diameters which seat in the frame, and an intermediate eccentric portion which carries the rocker. The left end is provided with a thumb piece for operation, and the right end is provided with a partial annular groove having two flat bottoms with which the short finger of the pivot plate engages to hold the same assembled to the frame, and also to yieldingly hold it in its "automatic" and "semi-automatic" positions.

The *safety* (31) consists of a round body with a thumb piece for operation and an annular groove with detents with which the long finger of the pivot plate engages to hold the safety assembled to the frame, and to yieldingly hold the safety in its "safe" and "fire" positions. The body portion of the safety forms a half-cylinder, so that in the "safe" position the safety engages with a recess in the sear and locks the same against movement, and in the "fire" position it is clear of the sear so that the same is free for movement.

The *sear* (32) is of rectangular shape provided at its rear end with an edge to engage the sear notch of the bolt. Its forward end is bifurcated to straddle the sear lever. It is pivoted approximately centrally. Its rear end is provided with a pocket for the sear spring, and with a transverse semi-cylindrical recess for the safety.

The *sear spring* (33) is seated in a pocket therefor in the sear and urges the sear to engage with the bolt.

The *sear lever* (34) is pivoted on the same pivot with the sear and fits between bifurcated ends thereof. It projects under one arm of the sear and forwardly beyond the sear for engagement with the disconnector. Its rear end is provided with a pocket for the sear lever spring.

The *sear lever spring* (35) fits into a pocket provided therefor in the sear lever and urges the sear lever to its normal inactive position.

The *trigger* (36) consists of a body portion provided with a pivot hole, a downwardly extending finger piece, a forwardly extending projection forming a pocket for the sear spring, and a rearward projection on which is mounted the disconnector.

The *trigger spring* (37) seats in a pocket provided therefor in the trigger and urges the trigger to its normal released position.

The *trip* (38) is mounted on the same pivot with the trigger, constructed to straddle the upper portion of the trigger. It has a forwardly extending projection to reach into the path of the magazine follower and a rear cross-bar to engage with an arm of the disconnector.

The parts described heretofore comprise the components of the Thompson Submachine Gun without magazines, stock or sights. These parts form individual attachments and are described as follows:

Magazines

The *Box Magazine* has a capacity for twenty cartridges in double column. It consists of a formed sheet metal body with a dovetail projection on its rear edge for engagement with the frame and a hole therein for engagement with the magazine catch. This rearwardly extending dovetail also furnishes a path for a rear projection on the follower for contacting with the trip to cause the bolt to be held open when the last shot is fired from the magazine.

The follower serves as a table for the cartridges in the magazine and is urged upward by the magazine spring, which is supported at the bottom of the magazine by the floor plate. This magazine consists of the four parts mentioned: the body, follower, magazine spring and floor plate.

The *Drum Magazines, Type L and Type C*, consist of circular pan-like bodies provided with covers. Rotatably mounted on a hub within the magazine is a rotor to which is attached a spring case housing a motor spring. To a flange on the hub is attached a ratchet consisting of a circular disc, so cut that four fingers are formed thereon and given a curvature to project beyond the plane of the ratchet. These fingers of the ratchet engage with inwardly protruding edges formed in the body of the magazine in such a manner that the fingers of the ratchet will pass over the projections when the hub is rotated in winding direction, and will abut on these projections when the hub is turned in opposite direction. The motor spring is fastened on its inner end to the hub, and on its outer end to the spring case.

The relation of these parts is such that when there are cartridges in the magazine, so that the rotor is not free to rotate, and the hub is turned in a feeding direction, the motor spring will become wound up and the fingers on the ratchet engaging the projections on the body will hold the hub from unwinding. The energy of the motor spring is then exerted through the spring case to the rotor to urge the cartridges to the mouth of the magazine.

Both the body and the cover of the magazine are provided with guide strips which form a spiral path for a train of cartridges placed axially with the hub. The rotor is provided with radial fingers which occupy the space in the magazine between the guide strips of the body and the cover. These fingers are so constructed that they intercept the spiral train of cartridges in such a manner that all the sectors are of equal length; that is, the distance between fingers on the inner row of the spiral train is the same as on the outer row of the spiral train. In this manner a group of cartridges between fingers on the inner spiral train occupy the same space between fingers on the outer row of spiral train; the cartridges throughout the train being at all times held compactly to prevent them from tumbling. In the larger magazine of the two (Type C) the fingers of the rotor at their extreme end are of such width that a pocket is provided therein for a cartridge in the outer path of the spiral train.

The magazine body and cover are provided with an opening which forms the mouth of the magazine; the train of cartridges is here arrested by deflectors which guide each cartridge upward into the mouth ready for feeding into the chamber. The body and cover are also provided with side plates secured on the outside thereof on opposite sides which engage guideways therefor in the receiver for attaching the magazine thereto. The side plate on the magazine body is provided at the bottom with a notch for engagement with the magazine catch to hold the magazine in position on the gun. The hub is secured to the magazine body with a spring clip. The cover is secured by a winding key which fastens to the end of the hub and by means of which the motor spring is wound.

The drum magazines are of two size: Magazine, Type L, of a 50-cartridge capacity, and Magazine, Type C, of a 100-cartridge capacity.

Sights

For use in connection with the butt stock to fire the gun from the shoulder, sights are provided.

The *front sight* consists of a one-piece member secured to the front end of the barrel with a pin.

The *rear sight* consists of a base which is riveted to the top of the receiver. The base is provided with side walls to protect the sight. It has a central housing for the plunger and plunger spring, and at its rear end it carries the sight leaf pivotally mounted thereon. The pivot end of the leaf is cylindrical and is provided with "V" slots for engagement with the plunger under pressure of the plunger spring. These "V" slots are so located that the plunger engages therewith when the leaf is in the "up" or in the "down" position and serves as a detent to hold the leaf in these two positions.

On the leaf is mounted a slide which is held in position by a small studded leaf spring secured on the edge thereof, the stud engaging with serrations on the edge of the leaf. Projecting from the slide is a forward wall notched for a battle sight and two lateral lugs which support a windage screw. This windage screw carries the eyepiece which extends below the slide. The eyepiece has an opening for a field view and an aperture for sighting. The upper end of the eyepiece is provided with a suitable edge for carrying an index mark for windage adjustment. Immediately above this edge on the battle sight wall is the windage graduation.

Butt Stock

To fire the gun from the shoulder there is provided a butt stock which can be readily attached to the frame, a catch locking the same in place.

The butt stock consists of a black walnut body. A slide which fits the frame of the gun is secured to the front end of the body by means of two screws. On this slide is mounted the catch to lock the butt stock to the frame.

At the butt end the stock is provided with a cavity for an oil can, and a butt plate is secured thereon with two wood screws. The butt plate is provided with a small circular hinged cap which serves as a door to the oil can cavity in the stock.

II. General Description of Gun and Operating Principle

The gun is composed of two distinct groups: the *Receiver* with the parts attached thereto and contained therein, and the *Frame* with its attached and contained parts.

The receiver forms the skeleton of the gun and has the barrel screwed thereon. Immediately beneath the barrel, at the front end of the receiver, is anchored the fore grip mount, to which is secured the fore grip. The fore grip mount is held in place by the barrel.

The receiver immediately to the rear of the barrel is provided with an opening for magazines and with a bolt well. Beyond the magazine opening is a uniform enlarged cavity which contains the body of the bolt, with recoiling space for the same, also the recoil spring and buffer.

The bolt consists of a rectangular body portion which fits into the cavity in the receiver, and a round forwardly projecting portion which fits into the forward bolt well of the receiver. The forward end of the bolt being of a reduced size permits the magazine to be brought up in close proximity to the chamber and as near as possible to the axis of the bore.

The lock consists of an "H" member with lugs extending on each side thereof. The sides of the lock engage with the bolt, and the lugs thereon engage with grooves in the receiver. The central bar is engaged by the actuator for control by the recoil spring and for manual operation.

The surfaces between the bolt and lock are at an angle of 70 degrees to the axis of the gun, and the surfaces between the lock and receiver at an angle of 45 degrees to the axis, so that there is formed on the lock an intercepted angle of 25 degrees between the bolt bearing and receiver bearing surfaces of the lock.

The explosive pressure of the cartridge is transmitted through the forward end of the bolt to the lock, and through the lock to the locking surfaces of the receiver. The bolt is limited in its forward travel by the abutment on the forward end of the rectangular section of the bolt abutting against the corresponding abutment in the receiver. In the portion of the bolt between this abutment and the lock is pivoted a hammer in the form of a reversed lever. The lower end of this hammer strikes the abutment in the receiver slightly in advance of the bolt. The upper end of the lever contacts with the firing pin which extends through the round forward portion of the bolt to the cartridge seat and is held in rearmost position by the firing pin spring. The hammer is so constructed that it will strike the abutment of the receiver and cause the discharge of the cartridge only when the bolt is completely locked.

The actuator rests slidably on top of the bolt and engages the cross-bar of the lock with two fingers; the rear finger also engages the forward end of the recoil spring. The cavity in the rectangular portion of the bolt forms a housing for the recoil spring, and the buffer serves as an arbor therefor. The buffer consists of a cylindrical tube closed at the forward end and having a flange at its rear end. This flange furnishes a seat for the rear end of the recoil spring and a buffing abutment for the rear end of the bolt. The buffer is mounted on the buffer pilot; a column of fiber discs is interposed between the forward end of the buffer and the buffer pilot, so that all pressure or shock on the buffer is transmitted to the buffer pilot through these buffing discs. The buffer pilot rests against the rear end of the receiver fitting loosely in a hole therefor, being held in place by a projection of reduced diameter.

Held in place by the buffer pilot and extending forwardly in undercuts in the receiver is a breech oiler formed of spring steel which holds oil saturated felt pads to relubricate the locking lugs on the lock at each recoil of the bolt. These pads also tend to keep the sides of the bolt lubricated.

The principle of bolt action is this, that during the period of high chamber pressure the lock is fixed in position by adhesion of its surfaces and moves to clear the locking surfaces in the receiver only after the high pressure in the chamber has subsided. The angle of the lock is so chosen that at the moment the lock is moved clear of the receiver locking surfaces, there is only sufficient residual powder pressure in the chamber to blow the cartridge case and the bolt rearwardly, eject the empty case and impart sufficient inertia to the bolt to completely compress the recoil spring and prepare the bolt for a new cycle of operation. On its forward movement, under impulse of the recoil spring, the bolt feeds a cartridge from the magazine into the chamber and as the bolt approaches its foremost position, the lugs on the lock engage the forward surfaces of the receiver locking grooves, which in conjunction with the recoil spring pressure on the lock through the actuator, drives the lock downwardly into locked position.

The receiver is provided on the right side with an ejection opening in a plane 30 degrees above the horizontal. In this same plane the forward end of the bolt is provided with an extractor secured by an undercut and limited in horizonal movement by a stud. The extractor is retained in its assembled position under its own spring tension.

The opposite side of the receiver is provided with an ejector which is screwed into place and secured in position by a projection on the end of a spring leaf engaging with a detent on the receiver. The ejector extends into the path of the bolt, a clearance cut on the bolt being provided therefor.

At the rear end the receiver is provided with a projection which contains the frame latch and frame latch spring. This member locks the frame to the receiver when the former is assembled thereto.

The parts thus far mentioned comprise the receiver group, which with the exception of the fore grip mount, with the fore grip and screw, and frame latch with its spring, composes the working parts of the gun.

The frame houses the entire trigger mechanism, furnishes a seat for the rear grip, an attachment for the box magazine, and contains a catch to hold the latter in place. The rear projection of the frame is also provided with a guideway, to which the butt stock can be attached.

Of greatest importance in the trigger mechanism group is the sear which under impulse of the sear spring engages one of the sear notches in the bolt when the latter is in retracted position. The trigger is mounted in the frame forwardly of the sear, and carries pivotally mounted on a rearward projection thereon the disconnector, which under impulse of the disconnector spring is continually urged toward the sear. The forward portion of the trigger houses the trigger spring, which urges the trigger to its normal released position.

Surrounding the upper portion of the trigger and mounted on the same pivot, the trip extends forwardly into the path of the box magazine follower, and rearwardly over a forwardly projecting member of the disconnector. The relation of the trip to the disconnector is such that when the trigger is pulled, the trip (when the last shot has been fired from the box magazine) will be lifted by the follower of the magazine and will move the disconnector away from the sear lever.

Between the sear and the disconnector is interposed the sear lever mounted on the same pivot with the sear. The sear lever is urged downwardly by the sear lever spring. The disconnector when moving upwardly, by a pull on the trigger lifts the sear lever, and the sear lever in turn lifts the forward projection of the sear, causing the rear projection to become depressed and disengaged from the bolt. The bolt is then free to move forward under impulse of the recoil spring and will reciprocate in automatic action until the cartridges are exhausted, or the trigger is released, or the bolt is arrested in its retracted position by the action of the follower of the box magazine on the trip when the last cartridge has been fed from the box magazine. In this last-mentioned instance the disconnector will be disconnected from the sear lever, which will be free to resume its normal position and will in turn allow the sear to re-engage the bolt.

Between the trigger and sear there is provided a rocker pivot which carries a rocker mounted on an eccentric portion thereof. This rocker extends upwardly when the rocker pivot is set for "automatic" position to within a very short distance of the bolt. The forward edge of the rocker rests against the disconnector. When the rocker pivot is rotated through an arc of 180-degrees to the "semi-automatic" position, the eccentric axis of the rocker pivot causes the rocker to project upwardly into the path of the bolt. There is provided on the bottom of the bolt a groove, to clear the rocker, of such length that the rear wall of this groove will strike the rocker at the final stage of the forward movement of the bolt, imparting to the rocker sufficient movement to cause it to throw the disconnector forwardly and disengage same from the sear lever. The sear lever immediately under impulse of the sear lever spring assumes its normal position, leaving the sear free to engage the bolt in retracted position.

The function of the sear lever is this: That as the bolt has reached its forward position and has caused the rocker to disengage the disconnector from the sear lever, if the sear lever were not free to move downward, the disconnector might re-engage the sear when the bolt starts rearwardly on its recoil, and the sear would then not be free to re-engage the bolt.

The position of the rocker controls the nature of fire of the gun, whether automatic or semi-automatic. *The rocker pivot can be turned from "automatic" to "semi-automatic" position, only when bolt is retracted.*

The safety is mounted at the rear end of the sear. It consists of a cylindrical body with its central portion halved to clear the sear. The sear is provided at the rear end with a half hole to engage the safety. When the safety is turned to "safe" position, it engages the sear so that the latter is positively blocked against movement. When the safety is turned to "fire" position, the half section of its body is rotated out of engagement with the sear and the sear is free to move.

The safety can be turned to "safe" position only when the sear is in engagement with the bolt in its retracted position. Obviously, since this is an open chamber gun, that is, the bolt when released by the trigger pull loads and immediately fires the cartridge, the gun with the bolt resting in forward position is completely inactive.

The pivot pin for the trigger and trip, and the pivot pin for the sear and sear lever are both secured to a spring plate, which is provided with projecting spring fingers engaging grooves on the safety and the rocker pivot, retaining these parts in assembled position.

The forward end of the frame is provided with a dovetail cut, which engages a corresponding dovetail member of the box magazine. On the side of the frame is pivoted the magazine catch urged into engagement with the magazine by a torsion spring. This magazine catch engages the box magazine by a stud extending forwardly through the center of the trigger guard; and also holds the drum magazines in position by the engagement of its forward end with a notch on these magazines.

The rear grip is secured to the frame by means of the rear grip screw. The frame is assembled to the receiver by undercut ways engaging corresponding ways on the receiver. When the frame is in foremost position on the receiver the frame latch pocketed at the rear end of the receiver is free under impulse of the frame latch spring to project downwardly into engagement with the frame, locking same to receiver.

There are three types of magazines that can be used on the gun, a box magazine and two drum magazines.

The box magazine of 20-cartridge capacity is provided with a dovetail which slides into a corresponding dovetail groove at the forward end of the frame, and is there held in position by the magazine catch.

The two-drum magazines of 50-cartridge capacity (Type L) and 100-cartridge capacity (Type C) do not attach to the frame as does the box magazine. They are provided, however, with slides which fit into grooves therefor in the magazine opening of the receiver, and are held by the same magazine catch which holds the box magazine in position.

Disabling Gun

The gun can be disabled and made useless by removing and destroying the pivot plate (Part No. 24).

III. Directions for Operating Gun

The gun may be set for either "automatic" or "semi-automatic" fire. It is provided with sights and a detachable butt stock. Three types of magazines having capacities of 20, 50 and 100 cartridges, respectively, may be used.

It is possible to fire the Submachine gun from the hip either automatically or semi-automatically, with any of the magazines, with or without the butt stock attached. Even though the rocker pivot is set for automatic fire, single shots may be obtained by a quick release of the trigger for each shot.

It is also possible to use the gun for aimed fire from the shoulder, either automatically or semi-automatcially, with any of the magazines.

The sight is graduated up to 600 yards for the 230-grain bullet ammunition, which is considered the effective range of the weapon, and the eyepiece thereof is laterally adjustable to correct for windage and drift. For rapid firing with the sight leaf down a 50-yard battle sight is provided.

Loading of Magazines

BOX MAGAZINES

The normal capacity of a box magazine is 20 cartridges. It may be possible to force an extra cartridge into the magazine, but this should be avoided.

The cartridges feed into the magazine with ease and without binding. If for any reason excessive force is required to feed the cartridges out of the magazine, the energy of the bolt is taxed to such an extent that a misfire may result. The forward edge of the magazine is rounded to prevent loading cartridges backward.

The lips of the mouth of the box magazine should be a distance of .55-inch apart. If by accident the magazine mouth should become deformed, the lips should be bent back to this dimension.

DRUM MAGAZINES.

To load a drum magazine, remove the winding key by lifting the flat spring thereon and sliding the key off. The cover can then be removed. Place the cartridges, bullet up, into the spiral track of the body, beginning with a full section at the mouth. The simplest method to begin loading is to fill one outer section and then rotate the rotor until this section reaches the mouth. Thereafter, continue to fill successive sections until the end of the spiral track has been reached. *Fill each section complete; do not skip any section and do not fill beyond the end of the spiral track.*

In order to obtain a capacity of 100 cartridges in the Type C magazine, a cartridge should be placed in each of the four fingers preceding the mouth.

After the magazine is properly filled replace the cover and key and wind to the number of clicks indicated on the magazine name-plate.

AUTOMATIC FIRE.

First cock or retract the bolt and move the safety to the "*safe*" position. Set rocker pivot to its forward position marked "automatic;" place a loaded magazine in position and assume a normal firing posture.

The matter of the firing posture for automatic firing is of great importance and should be carefully observed. The gun should be held with foregrip in the left hand and the rear grip in the right hand with index finger on the trigger. The rear end of the receiver should rest against the body in a natural position. The left foot should point in the direction of fire, and the right foot should be placed about 18 to 24 inches to the rear of the left foot and at an angle of from 60 to 90 degrees to the direction of fire. The knee of the left leg should be slightly bent so that the body leans forward. This leaning forward should be augmented for automatic fire from the shoulder.

There is no severe recoil to the gun, but the accumulation of rearward thrusts transmitted through the gun to the body will tend to push the operator backward if he is not well braced for the thrust, and will consequently interfere with the accuracy of fire.

There is also a tendency on the part of the gun to turn the operator on the axis of his body, unless he holds the gun close to himself and has his right foot braced as indicated above.

In automatic action the gun will fire continuously until the trigger is released, or until the magazine is emptied of its cartridges.

When firing from the box magazine the bolt will automatically be held in open position when the last shot has been fired therefrom. In both drum magazines the bolt will close on the empty chamber after the last shot has been fired. A rattling noise can be heard in these magazines which will indicate that they are empty. This noise is caused by the overtravel of the rotor due to an excessive magazine spring energy spending itself through the ratchet.

When removing a drum magazine, retract the bolt. The drum magazines can be placed into position or out of position only with the bolt retracted.

As a precautionary measure, it is deemed advisable to habitually set the gun at "*safe*" while changing magazines and during lulls in intermittent fire.

SEMI-AUTOMATIC FIRE.

First retract the bolt and move the safety to "*safe*" position. Set the rocker pivot to its rearward position marked "semi-automatic." *The rocker pivot can be moved to "semi-automatic" position only when the bolt is retracted.* Attach the butt stock to the gun; place a loaded magazine in position and assume a normal firing posture.

The *sight* is set by raising the leaf and sliding the slide to the range desired. Lateral correction is obtained by turning the small thumb screw.

In semi-automatic fire, it is also deemed advisable to habitually set the gun at "*safe*," except when actually firing.

MALFUNCTIONS.

In case of a misfire, due to faulty ammunition or otherwise, retract the bolt with a sharp quick pull on the actuator knob. This will insure ejection of the misfired cartridge.

In case of any other malfunction, retract the bolt and clear the throat and chamber of gun. If necessary, the magazine can be quickly detached for this purpose. While manipulating the gun, under these circumstances, it is deemed highly advisable to set the gun at "*safe*."

Precautions on Operating Gun.

(a) Each time before firing, reassure yourself as to the type of fire that is desired, whether "automatic" or "semi-automatic," and make certain that rocker pivot is set accordingly.

(b) See that magazine is loaded and assembled properly.

(c) Make it a practice to see that chamber is clear by looking through ejection opening.

(d) For anticipated action carry the gun with bolt cocked (retracted) and safety on. Otherwise bolt should be left in closed position on empty chamber, to relieve strain on recoil spring. To close bolt on empty chamber, *remove* magazine while letting bolt go forward slowly.

(e) For night operation, remember that with both the safety and rocker pivot in forward direction the gun is ready for automatic fire.

(f) Do not snap bolt on empty chamber unnecessarily.

IV. Directions for Dismounting and Assembling

TO DISMOUNT

1. REMOVE MAGAZINE.

Remove magazine by pressing upward with right thumb on magazine catch.

2. REMOVE FRAME FROM RECEIVER.

Turn safety to "fire" position and rocker pivot to "automatic" position. Pull the trigger and allow bolt to go forward gradually by retarding actuator with left hand.

Place gun upside down on knee or on a table, the barrel extending rearward, and steady against movement with the actuator knob. With thumb of left hand depress frame latch at rear end of the frame and with right hand tap frame sliding same rearward a short distance. Take the gun from table or knee, grasping rear grip in right hand, and grasping receiver with left hand *pull trigger* and slide frame off to the rear.

3. REMOVE RECOIL SPRING.

Support muzzle of barrel on table or knee, with under side of receiver facing operator. Grasp receiver with left hand, with thumb in position to engage the buffer. With thumb of right hand press down on buffer pilot which projects beyond end of receiver, and with thumb of left hand engage the flange of buffer. If the breech oiler follows, push same back with the fingers of right hand. Holding the buffer and pilot down with thumb of left hand grasp the end of buffer pilot with thumb and forefinger of right hand and withdraw this entire unit from the receiver.

Care should be taken to obtain a firm hold on the spring, buffer and pilot to prevent the recoil spring (same being compressed) from springing out of operator's hand.

4. REMOVE BOLT, LOCK AND ACTUATOR FROM RECEIVER.

Grasp receiver bottom up with left hand. Slide the bolt into rearmost position and withdraw.

Slide actuator with lock to foremost position and remove lock through inclined locking grooves in receiver.

Then again slide actuator to rearmost position and withdraw same.

5. REMOVE EJECTOR AND BREECH OILER FROM RECEIVER.

The ejector can be removed by lifting the leaf sufficiently to disengage the detent and unscrewing the same from receiver. The breech oiler can be removed by pressing its fingers together to clear undercut of the receiver. These two parts, however, need not be removed for ordinary cleaning purposes.

6. REMOVE SAFETY, ROCKER PIVOT AND ROCKER FROM FRAME.

Using the end of the actuator as a tool in the right hand depress the short finger of the pivot-plate and withdraw the rocker pivot; then remove the rocker.

Again using the actuator, but steadying the hand with thumb against trigger guard to prevent excessive movement, depress the long finger of the pivot plate and withdraw safety.

7. REMOVE PIVOT PLATE AND FIRING MECHANISM FROM FRAME.

Hold frame upright with the grip in right hand. Press simultaneously with both thumbs on the two pins of pivot plate. These pins project sufficiently far so that by a quick pressure thereon the body portion of the pivot plate will extend on the other side far enough to enable grasping same with fingers for withdrawal.

While withdrawing pivot plate with right hand, press down on the trigger and sear with left thumb to release pressure of springs on pivot pins to facilitate withdrawal.

The remaining components of the firing mechanism are then free to be removed. The disconnector can be removed from the trigger by simply withdrawing same.

8. REMOVE MAGAZINE CATCH FROM FRAME.

If required the magazine catch can be withdrawn from frame by rotating same counter-clockwise to its limit. Except for good reasons the magazine catch should not be removed, to avoid unnecessary straining of the magazine catch spring.

9. REMOVE FIRING PIN.

Drive hammer pin out of bolt from left side; the hammer, firing pin and firing pin spring will then tend to spring out under impulse of the firing pin spring. Caution should be exercised to prevent these parts from springing away and becoming lost.

10. REMOVE EXTRACTOR FROM BOLT.

With a cartridge case or some other means lift the extractor head sufficiently for the stud to clear its seat and withdraw by pulling forward. Caution should be exercised not to lift the extractor head excessively to prevent setting.

TO ASSEMBLE

1. ASSEMBLE TRIGGER MECHANISM.

First see that magazine catch is in assembled position.

Assemble disconnector to trigger by depressing disconnector spring and sliding disconnector into place.

Place trigger, trip, sear and sear lever into their respective positions in frame, making sure that forward end of sear lever rests on top of the disconnector. To align these parts press downward with end of left thumb on trigger and with base of thumb on sear. Insert the pivot plate and to avoid binding apply gentle pressure with ball of right hand over entire pivot plate.

2. ASSEMBLE SAFETY, ROCKER AND ROCKER PIVOT.

Insert safety from left side of the frame. With the actuator in the right hand, steadying same carefully to avoid excessive movement, depress the long finger of pivot plate and push safety home. Turn safety to "fire" position.

Place the rocker in position in frame with flat side against sear lever. Insert rocker pivot from the left side of frame. With actuator depress the short finger of pivot plate and push rocker pivot home. Turn rocker pivot to "automatic" position.

3. ASSEMBLE EJECTOR TO RECEIVER.

Screw ejector into receiver until stud on leaf engages and seats in depression therefor. *Do not screw or unscrew ejector while bolt is in closed position.*

4. ASSEMBLE EXTRACTOR TO BOLT.

Slide extractor into place, lifting head to clear stud.

5. ASSEMBLE FIRING PIN.

Place firing pin into firing pin spring and slide same into front end of bolt. Place the hammer in position, with rounded edge upward, and drive hammer pin into place.

6. ASSEMBLE ACTUATOR, LOCK AND BOLT TO RECEIVER.

Grasp receiver, bottom up, with left hand and insert actuator with knob to the front. Slide actuator to its foremost position. Introduce lock by engaging the lugs thereon in the locking grooves of the receiver, taking care that the arrow on the cross-bar of the lock is pointing toward muzzle of the gun, with word "up" reading correctly from rear.

Again slide the actuator with lock all the way to the rear and place the bolt into position.

7. ASSEMBLE RECOIL SPRING, BUFFER AND PILOT.

Slide bolt forward and rest muzzle of barrel on table or knee, grasp receiver with left hand and with right hand introduce recoil spring with buffer and buffer pilot. Push recoil spring down into bolt and let buffer pilot find its seat in receiver and snap into place.

8. MOUNT RECEIVER ON FRAME.

Grasp the frame with right hand in normal position, making sure that the safety is set at "fire" and the rocker pivot at "automatic". Slide frame on to receiver and at same time pull the trigger. The frame latch will lock the frame in position and the gun is now ready for action.

9. ATTACHING MAGAZINE.

The box magazine is attached by engaging the dovetail thereon with the dovetail groove in the forward end of the frame and moving magazine upward until caught by the magazine catch.

Drum magazines are introduced and removed from left side of the gun. They are held in place by the magazine catch. *When removing or attaching a drum magazine be sure that the bolt is retracted.*

Precautions on Dismounting and Assembling.

(a) See that bolt is forward and rocker pivot set at "automatic" before attempting to dismount frame from receiver. Rocker pivot must also be set at "automatic" and safety at "fire" when assembling frame to receiver.

(b) When assembling or removing extractor to or from bolt, do not lift extractor higher than necessary for lug to clear anchorage hole, to prevent setting or breaking of extractor.

(c) When assembling or removing ejector from receiver, make sure that bolt is not in closed position, as the ejector head engages with the ejector slot in front end of bolt.

Do not lift ejector leaf higher than necessary for disengaging stud with depression in receiver, to avoid setting or breaking of leaf.

(d) Do not remove pivot plate until frame has first been removed from receiver. If this precaution is disregarded serious difficulties may be entailed.

(e) When assembling or removing safety and rocker pivot, do not depress fingers on pivot plate more than necessary, to prevent setting or breaking.

(f) Do not remove breech oiler unless necessary.

(g) Do not remove magazine catch unless necessary. (The magazine catch can be assembled or removed only with the pivot plate partially withdrawn.)

V. Directions for Care and Preservation

Keep the gun well cleaned and oiled.

It is important that the gun be thoroughly cleaned after each day's firing regardless of the number of shots fired. Not only should the bore and chamber of the barrel be cleaned, but also all parts and all surfaces of the receiver, bolt, ejector and extractor as are contacted with the powder gasses.

For this purpose the frame should be removed from the receiver and the bolt should be taken out to thoroughly clean the front end thereof and the extractor. It may also be desirable to remove the extractor. The bolt-well and the throat of the receiver, as well as the ejector head, are readily accessible.

To prevent rusting as a result of the impregnation of powder gasses a saturated solution of sal soda water, consisting of one-quarter pound of sal soda per pint of water, can be used. Cloth patches soaked in the sal soda solution should be thrust through the bore with the cleaning rod, and the receiver and bolt surfaces affected should be wiped with a cloth saturated in the soda solution. Thereafter, all parts should be thoroughly dried and well oiled.

Instead of the soda solution for cleaning purposes, *gasoline* has been found to give good results.

In both cases, however, whether soda or gasoline is used, it is important that the parts be wiped dry after the wash and thoroughly oiled.

With exception of the magazines, it is not necessary to give the remaining parts of the gun a thorough cleaning after each firing, unless many hundred shots have been fired and it is desired to remove all powder residue or other accumulated dirt. The magazines should be treated and kept clean similar to the gun parts affected by powder gasses.

The cleaning rod accompanying the gun should be used in cleaning the barrel. The tip of this cleaning rod is of such diameter that it will accommodate one or two patches of flannel two inches square. Brass wire brushes, which screw directly on to the cleaning rod, are also furnished for cleaning.

LUBRICATION.

Any good grade of light oil may be used for lubricating the gun.

To lubricate, the frame should be removed from the receiver. Oil should be dropped over the pivot points of the trigger and trip, and the sear and sear lever; also over the disconnector and rocker.

Holding the receiver in the left hand, open side up, the bolt should be slightly retracted and oil should be dropped on the locking lugs of the lock, on the sides of the lock and on all sliding surfaces of the bolt and receiver.

The felt pads in the breech oiler should be kept well saturated with oil.

After reassembling the frame to the receiver the bolt should be retracted and a little oil should be dropped on the rounded front end of the bolt. The actuator knob should be worked back and forth several times to insure penetration of the oil to all parts of the mechanism.

All sliding surfaces should be oiled frequently and freely to insure perfect functioning of the gun.

VI. Ammunition and Ballistics

This gun fires the caliber .45 automatic pistol ball cartridge, the same as used in the caliber .45 U. S. Automatic Service Pistol, Model of 1911.

The 230-grain bullet ammunition is the standard ammunition for the gun, but the 200-grain bullet ammunition will function equally as well and will give an increased velocity.

The two types of ammunition produce the following chamber pressures and velocities:

Ammunition	*Average Chamber Pressure Pounds per Square Inch*	*Muzzle Velocity Feet per Second*
230-Grain bullet..........	12,470	918
200-Grain bullet..........	10,310	945

The sight is graduted for the 230-grain bullet ammunition.

When firing the 200-grain bullet ammunition the following sight settings should be used:

For 100 yards, set at 80 yards.
200 yards, set at 175 yards.
300 yards, set at 270 yards.
400 yards, set at 365 yards.
500 yards, set at 460 yards.

Due to the large caliber of the bore, the velocity of the ammunition is greatly affected by temperature. The sight was calibrated at a temperature of 35 to 40 degrees Fahrenheit. Adjustment of the sight setting may be necessary to correct for temperature.

Following are tables giving the height of the trajectory of the

230-grain bullet ammunition, and of the 200-grain bullet ammunition. There are also tables giving the drift, which is to the right.

The windage graduation is such that one unit of graduation will give a correction of one foot on a one-hundred-yard target.

HEIGHT OF TRAJECTORY ABOVE LINE OF SIGHT.

	RANGE Yards	*Height of Trajectory at Points Indicated*									
		50 Yds.	*100 Yds.*	*150 Yds.*	*200 Yds.*	*250 Yds.*	*300 Yds.*	*350 Yds.*	*400 Yds.*	*450 Yds.*	*500 Yds.*
		Ft.	Ft.	Ft.	Ft.	Ft.	Ft.	Ft.	Ft.	Ft.	Ft.
230-grain bullet	100	.40	0								
	200	1.40	1.90	1.55	0						
	300	2.55	4.25	5.00	4.60	2.95	0				
	400	3.95	7.00	9.20	10.12	9.80	8.20	4.90	0		
	500	5.58	10.20	14.00	16.50	17.70	17.60	15.86	12.50	7.15	0
200-grain bullet	100	.40	0								
	200	1.22	1.65	1.35	0						
	300	2.30	3.75	4.40	4.08	2.72	0				
	400	3.50	6.25	8.12	9.05	8.90	7.35	4.52	0		
	500	4.90	9.05	12.30	14.55	15.75	15.58	14.03	10.85	6.00	0

TABLE OF DRIFT.

AMMUNITION	*Drift to Right of Line of Bore of Gun at Points Indicated*									
	50 Yds.	*100 Yds.*	*150 Yds.*	*200 Yds.*	*250 Yds.*	*300 Yds.*	*350 Yds.*	*400 Yds.*	*450 Yds.*	*500 Yds.*
	Ft.	Ft.	Ft.	Ft.	Ft.	Ft.	Ft.	Ft.	Ft.	Ft.
230-grain bullet..	0	.10	.25	.50	.85	1.40	2.30	3.25	4.50	5.90
200-grain bullet..	0	.15	.35	.75	1.35	2.15	3.15	4.35	6.10	8.05

TABLE OF PENETRATION.

The remaining energy of the bullet at the various ranges is shown in the following table in which the figures show the number of ¾ inch yellow pine boards spaced one inch apart, that were penetrated by bullets from the firing points indicated.

Ammunition	*Number of boards penetrated from firing points indicated.*					
	Point Blank	*100 Yards*	*200 Yards*	*300 Yards*	*400 Yards*	*500 Yards*
230-Grain bullet....	6¾	6	5¼	4½	4	3¾
200-Grain bullet....	6¾	5¾	5	4¼	3¾	3¼

VII. Data

1.—Table of Weights.

	Pounds
Gun complete with sights (without magazine or butt stock).....	8.5
Five box magazines (empty).....	2.0
One Type L drum magazine (50-cartridge), empty.....	2.5
One Type C drum magazine (100-cartridge), empty.....	3.9
One butt stock.....	1.5
One web gun case.....	1.9
100 Cartridges (230-grain bullet).....	4.6
2000 Cartridges (230-grain bullet), packed.....	110.0
Trigger pull.....	8 to 10

2.—Table of Dimensions.

	Inches
Overall length of gun without stock.....	23.2
Overall length of gun with stock.....	31.8
Overall length of barrel.....	10.5
Length of bore.....	9.76
Rifling, right hand one turn in 16 inches.	
Heel of butt stock below top of receiver.....	3.3
Heel of butt stock beyond rear of receiver.....	8.6

3.—Magazines.

Type XX Box Magazine (20 cartridge)
  1.0 by 1.7 by 6.2 inch.
Type L Drum Magazine (50-Cartridge)

Outside diameter	6.65 inch
Thickness of body	1.40 inch
Overall thickness	2.17 inch

Type C Drum Magazine (100-Cartridge)

Outside diameter	8.63 inch
Thickness of body	1.40 inch
Overall thickness	2.17 inch

4.—Sight.

Height of front sight above axis of bore	1.04 inch
Sight radius	22.30 inches
Battle sight range	50 yards

5.—Cartridges.

Length of bullet:	
230-Grain bullet	.662 inch
200-Grain bullet	.582 inch
Length of case	.895 inch
Overall length of cartridge:	
230-Grain bullet	1.265 inch
200-Grain bullet	1.258 inch
Diameter of bullet (cylindrical portion)	.450 inch
Diameter of cartridge	.472 inch
Weight of bullet	200 and 230 grains
Weight of powder charge	5 grains
Weight of cartridge case and primer	89 grains
Weight of cartridge:	
230-Grain bullet	324 grains
200-Grain bullet	294 grains

APPENDIX B

AUTO-ORDNANCE CORPORATION JOBBERS NET PRICE LIST.
EFFECTIVE FEBRUARY 1, 1928. M1928 THOMPSON NOT YET LISTED.

	Cost to You	*Price to Dealer*	*Price to Consumer*
MODEL NO. 21 A—Thompson Submachine Gun, Standard Grade complete with one 20-cartridge capacity box magazine	$115.00	$135.00	$175.00
MODEL NO. 21 A C—Thompson Submachine Gun, Standard Grade, complete with one 20-cartridge capacity box magazine, and with Cutts Compensator fitted	135.00	157.50	200.00
MODEL NO. 27 S C—Anti-Bandit Gun. Thompson Semi-Automatic Carbine, Special Grade, with *horizontal foregrip,* complete with Cutts Compensator and with one 20-cartridge capacity box magazine	135.00	157.50	200.00
MODEL NO. 27 S C D—Anti-Bandit Gun. Thompson Semi-Automatic Carbine, Special Grade, with *horizontal foregrip,* complete with Cutts Compensator, with one 50-cartridge capacity drum magazine and one 20-cartridge capacity box magazine	147.60	173.25	221.00
MODEL NO. 27 S C D S—Anti-Bandit Gun. Thompson Semi-Automatic Carbine, Special Grade, with *horizontal foregrip,* complete with Cutts Compensator and with swivels and web adjustable sling straps fitted, with one 50-cartridge capacity drum magazine and one 20-cartridge capacity box magazine	150.80	177.25	227.00
MODEL NO. 27 A—Thompson Semi-Automatic Carbine, Standard Grade, with *vertical foregrip,* complete with one 20-cartridge capacity box magazine	115.00	135.00	175.00
MODEL NO. 27 A C—Thompson Semi-Automatic Carbine, Standard Grade, with *vertical foregrip,* complete with one 20-cartridge capacity box magazine, and with Cutts Compensator fitted	135.00	157.50	200.00
TYPE X X—20-cartridge capacity box magazine	1.80	2.25	3.00
TYPE L—50-cartridge capacity drum magazine	12.60	15.75	21.00
Special Peters-Thompson Shot Magazine—18-cartridge capacity—for shot cartridges	1.80	2.25	3.00
Brass cleaning rod with removable wire brush tip (like rod furnished free with each gun)	.90	1.15	1.50
Short Breech Cleaning Brush (wire handle with bristles)	.45	.55	.75
Web Gun Carrier Case, Type A, with holster for stock and four pockets for 20-capacity box magazines (state whether for gun with compensator, or not)	9.90	12.40	16.50
Web Gun Carrier Case, Type B, for mounted use (state whether for gun with compensator, or not)	9.90	12.40	16.50
Web Belt Carrier Outfit, Type D, complete with four pocket case for 20-round box magazines and case for 50-round drum magazine	6.60	8.25	11.00
Belt (separately)	1.20	1.50	2.00
4-pocket case (separately)	3.00	3.75	5.00
50-drum case (separately)	3.00	3.75	5.00

Auto-Ordnance catalog and price list, 1929.

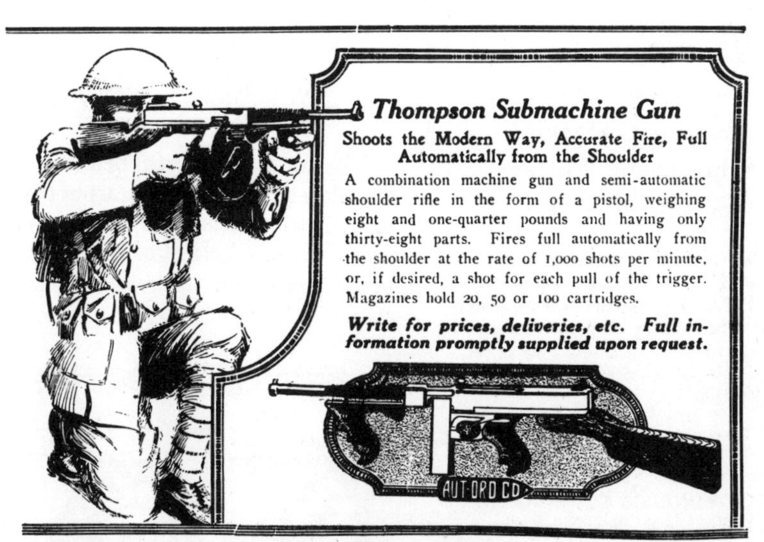

Army & Navy Journal, 1920-21.

A Sure Defence

Against

Organized Bandits and Criminals

THE epidemic of attacks by organized bands of criminals throughout the country emphasizes the need of additional protection for financial and industrial centers and for resident communities. The old time safeguards are inadequate to foil the carefully planned raids of heavily armed bandits, whose "getaway" is assured by high powered automobiles. Improved and more powerful methods of defense have become necessary.

The Thompson Anti-Bandit Gun has been placed upon the market to meet these conditions. It is the most effective portable firearm in existence, automatically firing at the rate of 1,000 shots a minute, or single shots as fast as the trigger can be pulled. One man armed with a Thompson has the defensive power of a dozen men armed with older type weapons.

Specifications — Colt manufacture, caliber .45, weight 8½ lbs., length 22 inches, ammunition, Colt automatic pistol ball (birdshot for short ranges). Magazines hold 20, 50 or 100 cartridges.

The gun is a powerful deterrent. It strikes terror into the heart of the most hardened and daring criminal. The moral effect of its known possession is an insurance of its own.

Introducing

on the side of Law and Order The Thompson Anti-Bandit Gun, as an auxiliary to the police and other safeguards, for the better protection of

Suburban Estates
Industrial Plants
Stores
Banks
Railway Yards
Express Companies
Shipping and Wharves
and other properties.

It is Government tested and now adopted for protective purposes by the Police Departments of more than fifty cities and by the Constabularies of many States. It is sold only to responsible parties after a thorough investigation.

For descriptive literature and prices
write or telephone
Department A.

AUTO-ORDNANCE CORPORATION
302 Broadway, New York City
Tel. Worth 3993

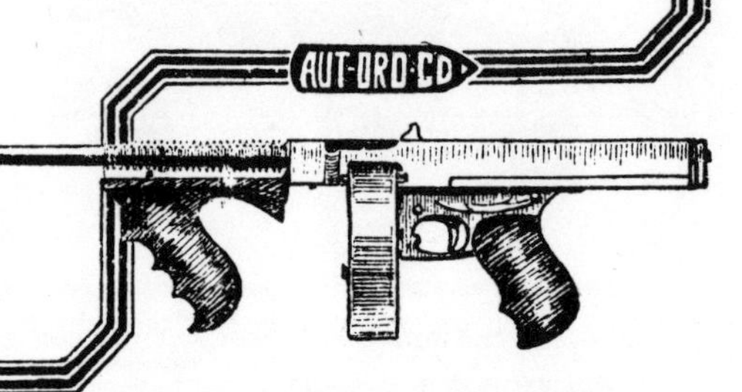

New York Herald, Jan. 31, 1922.

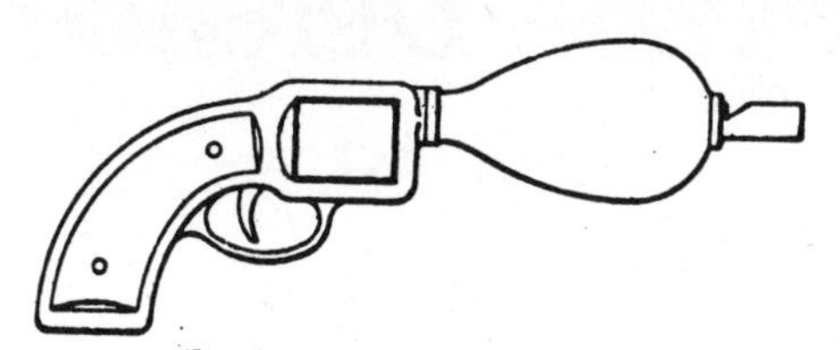

Patent drawings of John Thompson's signal gun and tubular airplane.

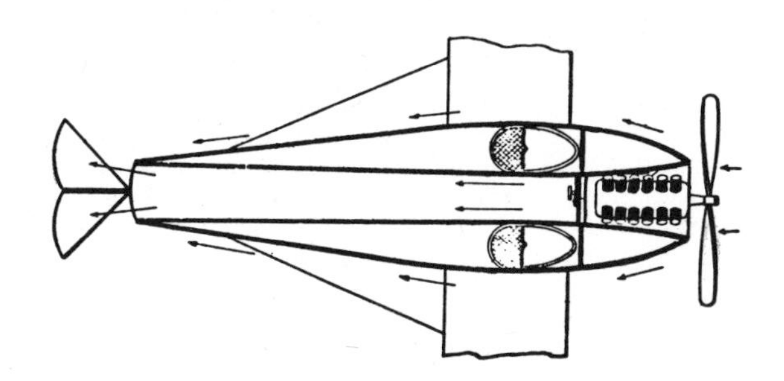

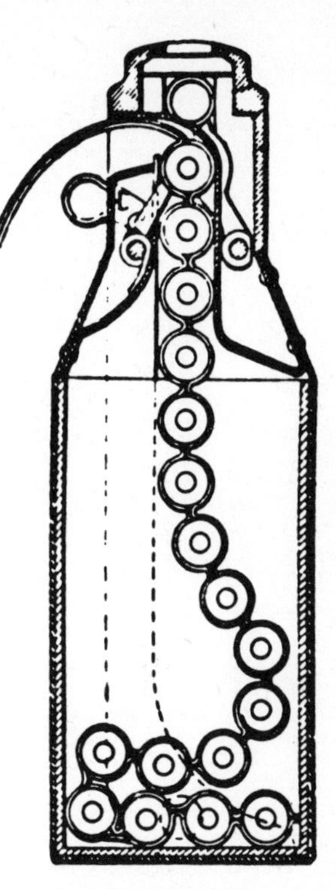

Ammunition belt container for the "Persuader."

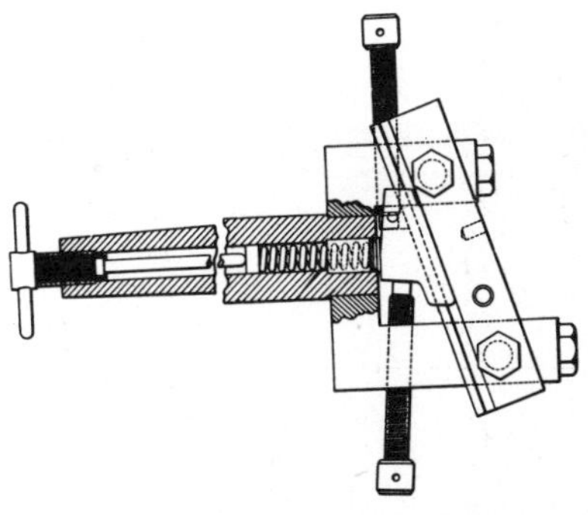

Patent drawing of Blish demonstration mechanism.

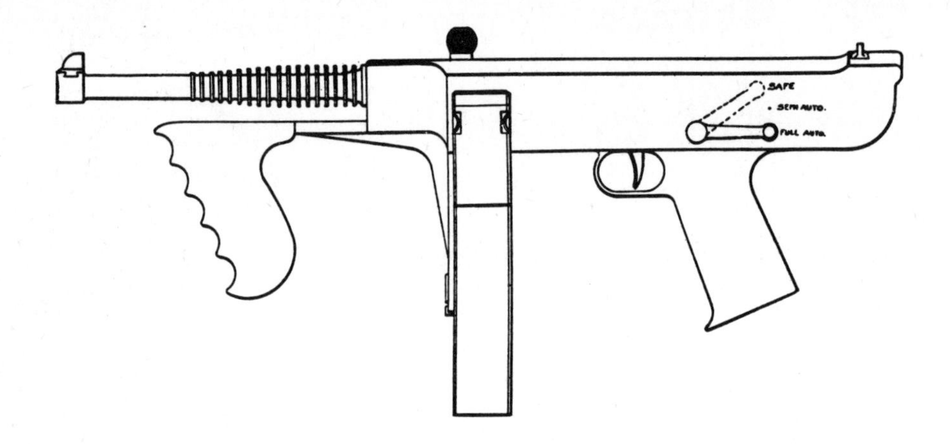

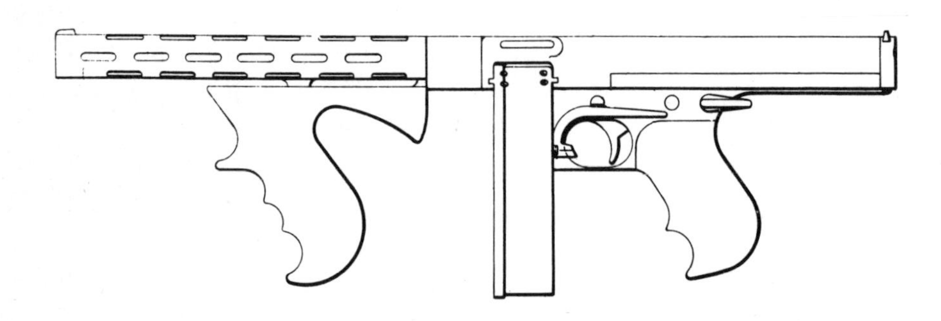

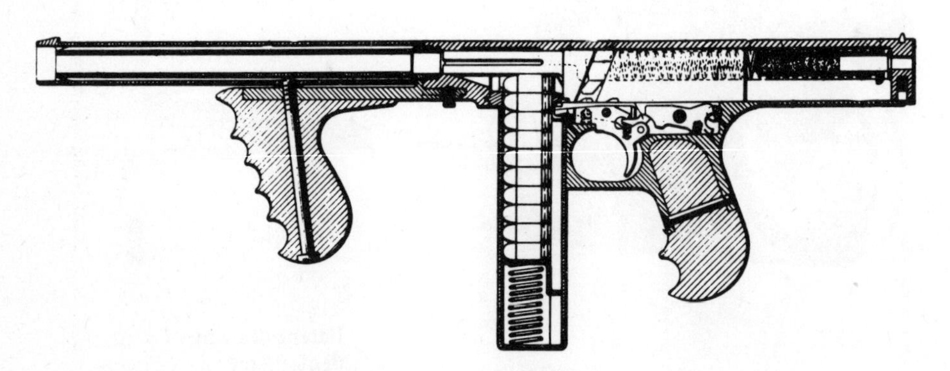

Thompson patent drawings
showing early design variations.

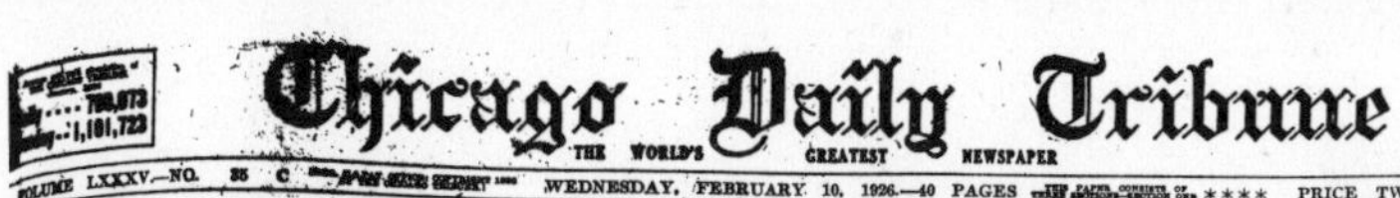

Chicago Daily Tribune

THE WORLD'S GREATEST NEWSPAPER

FINAL EDITION

VOLUME LXXXV.—NO. 35 C WEDNESDAY, FEBRUARY 10, 1926.—40 PAGES PRICE TWO CENTS

MACHINE GUN GANG SHOOTS 2

Italy Seizes Tyrol Arms Cache; 50 Held

CHARGE REVOLT PLOT STIRRED BY BAVARIANS

Reichstag Warns Mussolini.

BULLETIN.

ROMA, Italy, Feb. 9.—(P)— persons have been arrested ... of arms and ammunition have been seized by Fascist ... near Lavarone in the Austrian territory of Trentino (South Tyrol). The Fascist ... Bavarians were ... Italy.

NEWS SUMMARY

DOMESTIC.

Downstate posse shoots it out with Chicago gunmen; kills one, wounds two; balks bandit rescue. Page 1.

Countess of Cathcart, held at Ellis Island because she is a divorcee, assails United States. Page 1.

Trade commission investigation of baking combine adjourns for two weeks. Page 14.

University of Wisconsin sends home students who fall in studies. Page 20.

FOREIGN.

Italians seize arms cache and arrest fifty in Tyrol, charging Bavarian revolt plot; reichstag warns Mussolini. Page 1.

Paris art students strike demanding more pulchritude and less age in their models. Page 1.

British coal commission recommends nationalization of mines. Page 2.

Stillmans reach sensible stage in their reconciliation; agree in quiet places. Page 10.

Premier Mussolini declares Italian-British accord serves as guarantee for Europe's peace.

TEMPTED

RAKE SALOON IN BEER WAR; ONE MAY DIE

Hangout Riddled by Steel Bullets.

(Pictures on back page.)

Thirty-seven bullets from a light automatic machine gun were poured into the saloon of Martin (Buff) Costello, 4127 South Halsted street, last night, by gangsters striving to assassinate two rivals for the highly profitable south side traffic in good beer.

Both men were wounded. William Wilson, 339 South Leavitt street, was shot in the head and probably fatally wounded. John ...

Find Bessner Guilty; Given 20 Year Term

BULLETIN.

The jury trying Roger Bessner for the murder of Bonifaciao Hernander in a holdup, after an all night session, reached a verdict at 6 o'clock this morning. Bessner was found guilty and his punishment fixed at 20 years in the penitentiary.

The jury debating the fate of Roger L. Bessner, who is charged with the murder of Bonifaciao Hernandes, chauffeur for Dr. Henry W. Gross, 330 Galt avenue, in a holdup, was reported deadlocked last midnight. The case was given to the jurors in the court of Judge Emanuel Eller early in the evening after Assistant State's Attorney Charles Mueller had made a dramatic appeal for the death penalty.

It was reported that the jurors were debating the possibility that Dr. Gross might have killed the chauffeur himself in the gun battle incident to the holdup. No ballots had been taken.

Widow Screams "Lie! Lie!"

... defense attorneys in ...

SMALL'S FINAL PLEA REJECTED; TALK OF OUSTER

Carlstrom to Decide on Removal Action.

With The Illinois Supreme court's complete and final refusal yesterday to reconsider its finding against him Gov. Len Small found himself face to face with these facts:

1 Within a short time he will be compelled to account for and pay over more than $1,000,000 in interest on public funds withheld during his second term as state treasurer.

2 Ouster proceedings will be begun against him on grounds that he never was legally elected to office, since he was ...

Chicago Tribune, Feb. 10, 1926.

Army & Navy Journal, Nov. 5, 1921.

Details of the Larsen All-Metal Attack Plane

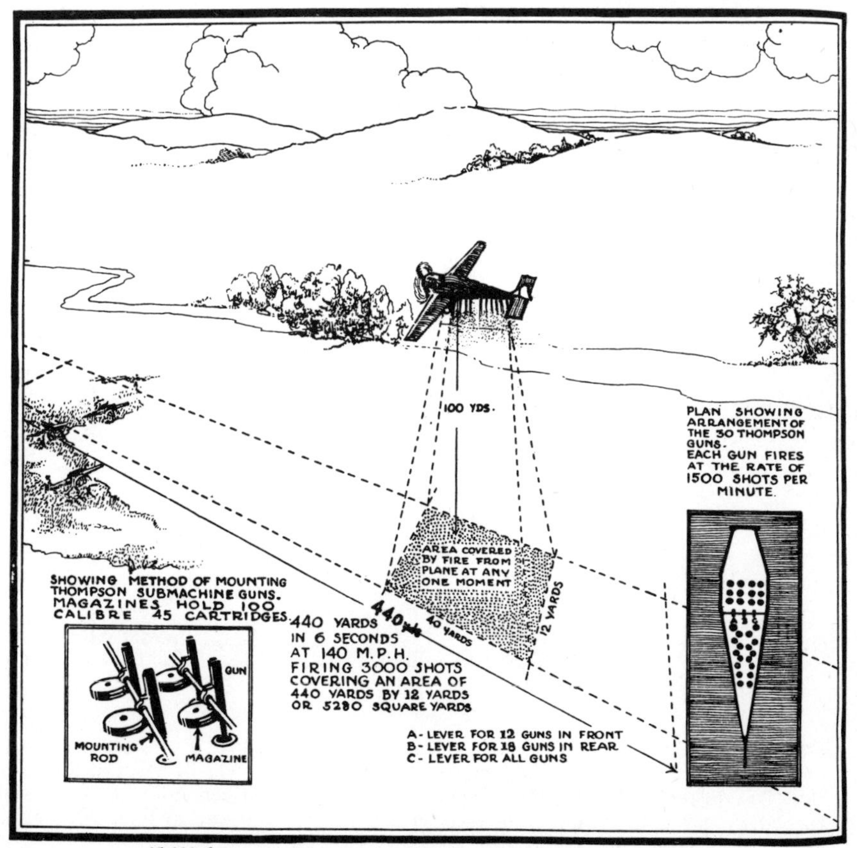

45,000 shots per minute can be fired from the the 30 automatic guns mounted on this plane while traveling at a speed of 140 miles per hour.

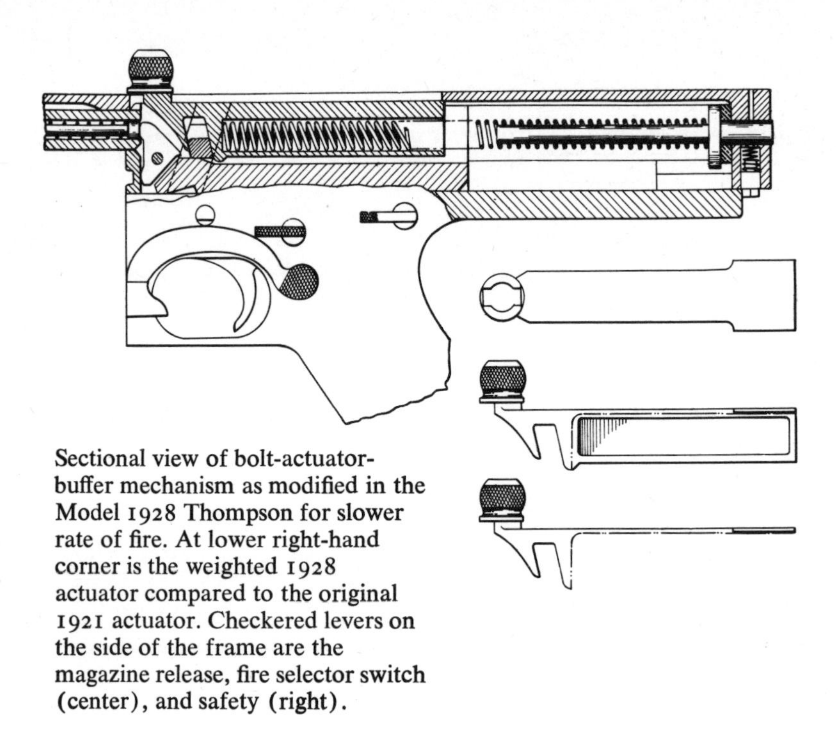

Sectional view of bolt-actuator-buffer mechanism as modified in the Model 1928 Thompson for slower rate of fire. At lower right-hand corner is the weighted 1928 actuator compared to the original 1921 actuator. Checkered levers on the side of the frame are the magazine release, fire selector switch (center), and safety (right).

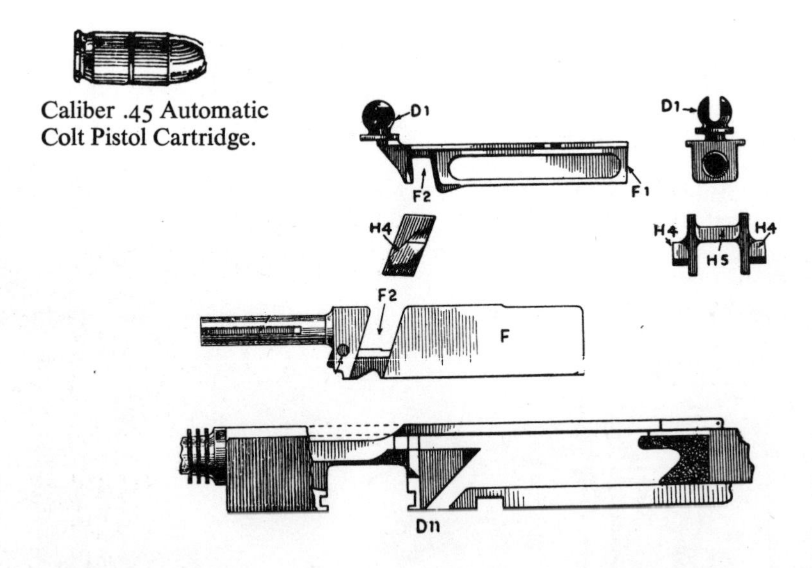

Caliber .45 Automatic Colt Pistol Cartridge.

D1—Cocking handle. D11—Slots in receiver wall. F—Bolt. F1—Actuator (1928 modification). F2—Locking slot, mates with H-piece, H5. H4—Lugs which engage slots D11 in receiver walls. H5—H-piece, rides in bolt and actuator slots (F2) with lugs (H4) engaging slots D11 in receiver walls; links actuator to bolt and "locks" against surfaces of slots D11.

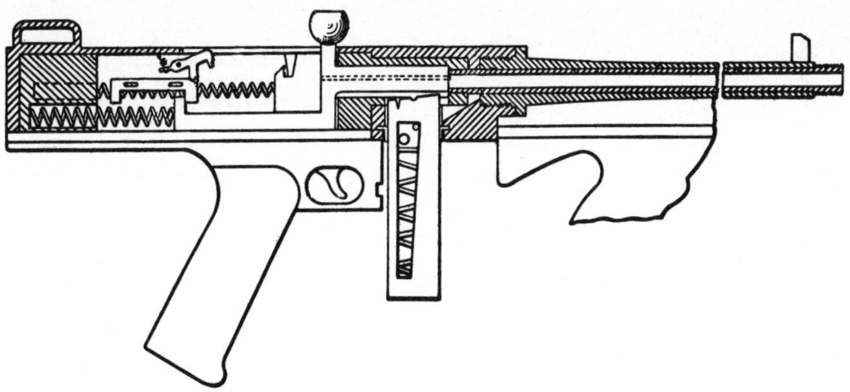

Patent drawing showing placement of Robbins .22-caliber conversion unit in Thompson receiver.

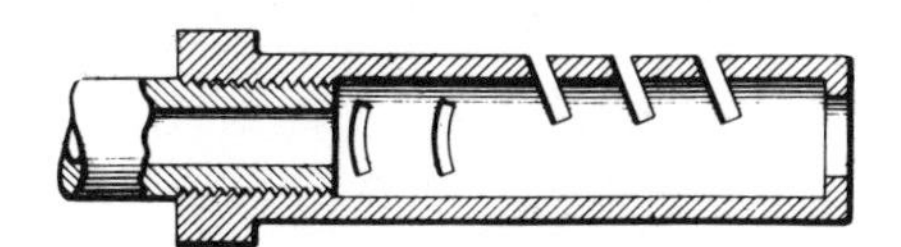

Patent drawing of Cutts compensator.

Original "Aut-Ord-Co" trademark was stamped on early Colt-made guns before it was replaced by the "Thompson" design around the end of 1921.

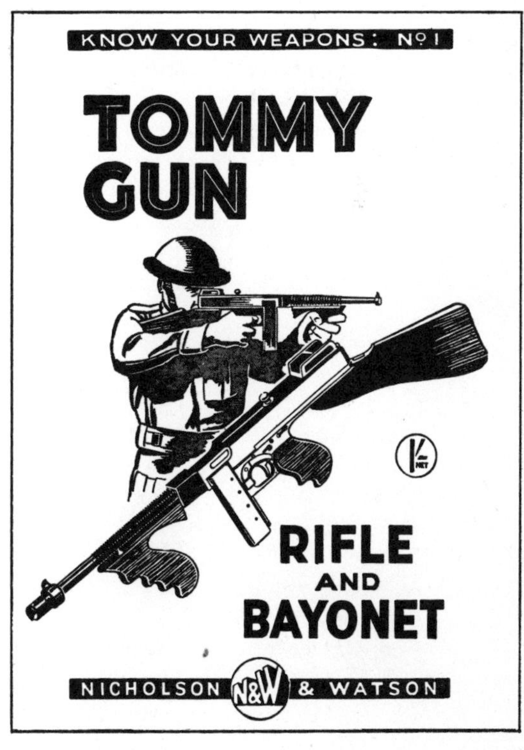

World War II British Manual

Index